CHEMISTRY RESEARCH AND APPLICATIONS

HETEROCYCLES

SYNTHESIS, REACTIONS AND APPLICATIONS

CHEMISTRY RESEARCH AND APPLICATIONS

Additional books and e-books in this series can be found on Nova's website under the Series tab.

CHEMISTRY RESEARCH AND APPLICATIONS

HETEROCYCLES

SYNTHESIS, REACTIONS AND APPLICATIONS

JONAS M. DANIELSEN
EDITOR

nova
science publishers
New York

Library of Congress Cataloging-in-Publication Data

ISBN: 978-1-53617-804-3
Library of Congress Control Number: 2020935951

Published by Nova Science Publishers, Inc. † New York

CONTENTS

PREFACE

Considering the extensive applications that have been discovered for naphthoquinone-triazole hybrids in the search for new bioactive molecules, *Heterocycles: Synthesis, Reactions and Applications* first comprehensively addresses recent synthetic strategies for their preparation.

The authors highlight recent developments in reaction strategies for the synthesis of dihydro-, tetrahydro-, pyrano-pyrans and tetrahydrofurans and their applications in natural product synthesis.

The important structural features of heterocyclic compounds are reviewed, as well as their deviation from the planar configuration in the form of puckering or buckling, and methods of suppression.

The closing study explores the reaction of azides with Grignard or organolithium reagents in an inter- or intramolecular fashion to form linear or cyclic triazene intermediates, respectively.

Chapter 1 - Naphthoquinones are a group of natural and synthetic substances that participate in various biochemical functions in plants and animals, as they are involved in several important biological processes. Their performance in such biochemical functions is directly related to the functional groups linked to the naphthoquinone system. The introduction of the 1,2,3-triazole ring into their structure allows these molecules to bind to a variety of enzymes and receptors in the biological systems through non-covalent interactions, leading to a wide variety of biological activities.

For instance, naphthoquinone-triazole hybrids have showed cytotoxicity against several cancer cell lines. As a consequence, several methods for preparing such scaffolds were reported recently, leading to wide molecular diversity and new classes of bioactive compounds. Considering the extensive applications that naphthoquinone-triazole hybrids have gained in the search for new bioactive molecules, this chapter will comprehensively address recent synthetic strategies for their preparation.

Chapter 2 - Heterocycles are the basic core structure in a plethora of naturally occurring bioactive molecules. Oxygen containing heterocyclic compounds such as dihydro-, tetrahydro-, pyrano-pyrans and tetrahydrofurans are the key building blocks of various natural and biologically active products. Methodologies such as Prins cyclization, oxonium-ene, hetero-Diels-Alder, metal catalyzed cyclization, 1,n-enyne rearrangement reactions are commonly practiced for the synthesis of above heterocycles. To mention a few, decytospolide A, which has tetrahydropyran as its core structure can be synthesized using Prins cyclization reaction and has antibacterial and antitumour properties. (+)-Muconin has both tetrahydro-pyran and –furan rings and shows potent *in vitro* cytotoxicity against PACA-2 (pancreatic cancer) and MCF-7 (breast cancer). Saikia and co-workers synthesized bioactive substituted tetrahydropyrans *via* oxonium-ene reaction and natural product (+)-civet *via* Prins cyclization reaction. In 2009, Krause and co-workers reportedly synthesized natural products (+)-linalool oxide, (-)-isocyclocapitelline and (-)-isochrysotricine using gold catalyst. As such, significant methodologies are reported for the synthesis of dihydropyrans and tetrahydropyrans separately over the years, but reports solely concentrated on oxygen heterocycles as a whole and their reaction methodologies and applications are not reported yet. This review will highlight recent developments (since 2014) in reaction strategies for the synthesis of dihydro-, tetrahydro-, pyrano-pyrans and tetrahydrofurans and their applications in natural product synthesis.

Chapter 3 - This chapter is devoted to one of the most important structural features of heterocyclic compounds that defines their properties, the deviation from the planar configuration in the form of puckering or buckling, and to the methods of its suppression. First, the authors show that all the distortions of the planar configuration have a common origin, the pseudo Jahn-Teller effect (PJTE), and the authors explain how this effect produces the atomic out-of-plane displacements in function of the systems' parameters. The latter include the ground and low-lying excited states of the undistorted configuration, the energy gaps between them, and the vibronic coupling constants. The authors demonstrate how these PJTE parameters can be evaluated by means of electronic structure calculations, using some hexa-heterocyclic compounds as examples. The knowledge of the main features and explicit values of the parameters of the PJTE open a way to directly influence this effect and its produced distortions of the planar configuration by means of specific, *targeted* external implications (perturbations) in the form of oxidation, reduction, excitation, chemical substitution, and coordination to outer atoms or atomic groups. All these methods were successfully realized. Illustrative examples include a series of carbon and silicon tetra- and hexa-cyclic compounds, and more complicated tricyclic and polycyclic systems like dithiine containing dithianon, thiantrene, anthracene, sumanenes, etc. as well as extended 2D systems including silicene, phosphorene, germanene, and graphitic carbon nitride.

Chapter 4 - Organic azides are among the most versatile functional groups in organic chemistry. In addition to the widely known traditional reactivity pathways of the azido group itself, the additional capability of azides to form key reactive intermediates in a straightforward manner greatly expands their versatility and synthetical utility. One of the most salient reactive intermediates that can be readily accessed from organic azides is the triazene group. This functional group consists of three contiguous nitrogen atoms with a double bond between the first and second nitrogen. These adducts were recently divided, by Bugarin, in three categories: linear, cyclic, and π-conjugated triazenes. Once these adducts/intermediates have been formed, they can be immediately transformed into a variety of functional groups. In the present chapter, we'll discuss the reaction of azides with Grignard or organolithium reagents either in an inter- or intramolecular fashion to form linear or cyclic triazene intermediates, respectively. Through these reactions, azides can be converted into important functionalities such as; amines, amides, and nitrogen heterocycles, which in turn can function as synthetic building blocks for more complex chemical structures. Additionally, the authors will present known applications that make use of the power of azides' reactivity for the synthesis of prominent chemotherapeutic drugs bearing a triazene core structure. Last but not least, throughout the latter part of the chapter, the authors will explore recently reported azide reactivity with *N*-heterocyclic carbenes under basic conditions, which allows their conversion into aldehydes, ketones, ethers, thioethers, (hetero)biaryls, and aryl amides. This novel reactivity permitted the conversion of organic azides into useful functional groups under mild, transition-metal-free conditions, for example; (hetero)biaryls and aryl amides formation under a choice of photochemical or sunlight irradiation.

In: Heterocycles
Editor: Jonas M. Danielsen
ISBN: 978-1-53617-804-3

Chapter 1

SYNTHESIS AND APPLICATIONS OF NAPHTHOQUINONE-1,2,3-TRIAZOLE HYBRIDS

Vitor Francisco Ferreira[1,*], PhD, Patricia Garcia Ferreira[1], PhD, Thais de Brito da Silva[1], Danielle Pagliaminuto Portella[1], Fernando de Carvalho da Silva[2], PhD, Luana da Silva Magalhaes Forezi[2], PhD, Carolina Guimarães de Souza Lima[2], PhD, Acácio Silva de Souza[2], PhD, Ruan Carlos Busquet Ribeiro[2], Fernanda Petzold Pauli[2] and Daniel Tadeu Gomes Gonzaga[3], PhD

[1]Faculdade de Farmácia, Universidade Federal Fluminense, Niterói, Brazil
[2]Instituto de Química, Universidade Federal Fluminense, Niterói, Brazil
[3]Universidade Estadual da Zona Oeste, Rio de Janeiro, Brazil

* Corresponding Author's E-mail: vitorferreira@id.uff.br.

ABSTRACT

Naphthoquinones are a group of natural and synthetic substances that participate in various biochemical functions in plants and animals, as they are involved in several important biological processes. Their performance in such biochemical functions is directly related to the functional groups linked to the naphthoquinone system. The introduction of the 1,2,3-triazole ring into their structure allows these molecules to bind to a variety of enzymes and receptors in the biological systems through non-covalent interactions, leading to a wide variety of biological activities. For instance, naphthoquinone-triazole hybrids have showed cytotoxicity against several cancer cell lines. As a consequence, several methods for preparing such scaffolds were reported recently, leading to wide molecular diversity and new classes of bioactive compounds. Considering the extensive applications that naphthoquinone-triazole hybrids have gained in the search for new bioactive molecules, this chapter will comprehensively address recent synthetic strategies for their preparation.

Keywords: hybrids, naphthoquinones, 1,2,3-triazole, synthesis

INTRODUCTION

Molecular Hybridization in Medicinal Chemistry

Molecular hybridization is an important tool for the design of novel bioactive substances in medicinal chemistry, having been widely employed in the search for new series of compounds with potential biological activity in recent times. According to E. K. Ariens, hybrid drugs can be defined as "agents which combine in one molecule two, in their mechanism of action (receptors) different types of biological action" [1]. In a more elaborate concept, Fraga et al. describe molecular hybridization as "a strategy of rational design of new ligands or prototypes based on the recognition of pharmacophoric sub-unities in the molecular structure of two or more known bioactive derivatives which, through the adequate fusion of these sub-unities, lead to the design of new hybrid architectures that maintain pre-selected characteristics of the original templates" [2].

With the significant advances in the field of bioreceptors, especially the discovery of multiple receptor subtypes, as well as in the knowledge of 3D target protein structures, research has started to focus on finding new and more selective target-based drugs and prototypes. Within this context, the development of hybrid compounds has become a valuable tool for maximizing the selectivity and biological activity of such molecules [3].

Molecular hybridization also appears to act beyond molecular levels, having major therapeutic benefits, e.g., the replacement of drug cocktails (combination therapy) in some cases [4, 5, 6]. Hybrid construction allows different drugs or groups of drugs to act synergistically or by amplifying their effects, as well as minimizing or neutralizing side effects by associating one portion of the hybrid with another (Scheme 1) [7, 8]. As a consequence, problems related to drug interactions when more than one drug are administered simultaneously, or even poor efficacy in combination therapy for diseases resistant to conventional therapies, can be overcome, resulting in an overall increased efficiency in the clinical treatments [9, 10].

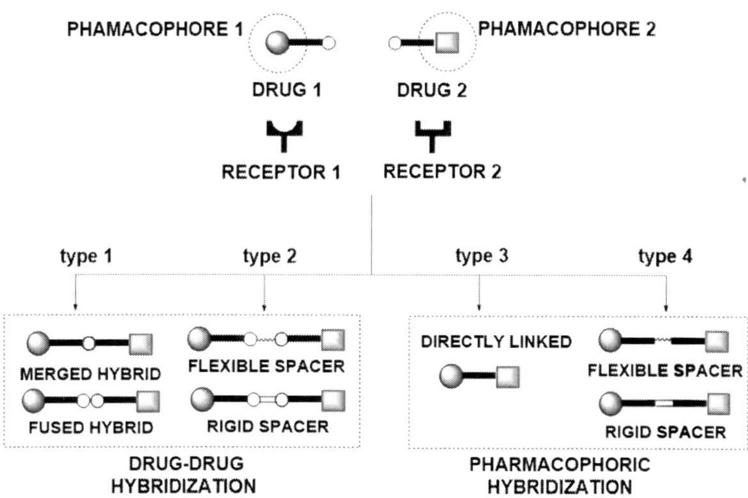

Scheme 1. Different molecular hybridization approaches.

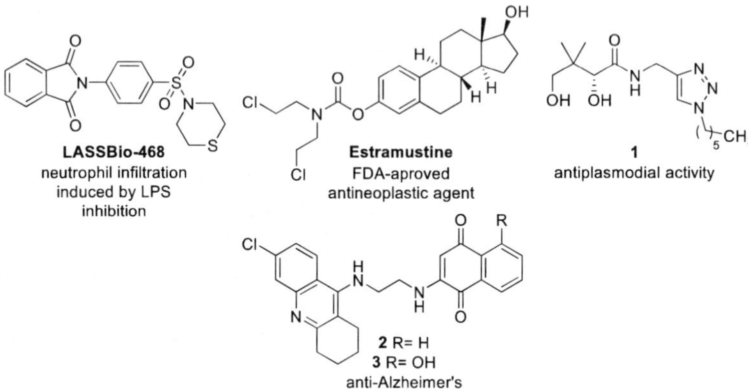

Figure 1. Structure of several classes of hybrid drugs and biologically active compounds.

Hybrid molecules can be designed and then synthesized from numerous sources, and owing to their great diversity, heterocycles are highly regarded in this field. The structural diversity of hybrid compounds is exemplified by the large number of molecules with biological activities currently found, either as drugs or drug prototypes (Figure 1).

An important example in which molecular hybridization has significantly improved the properties of a drug is the case of thalidomide. This compound is known for its antiemetic and sedative effects, but also for being teratogenic, a classical example in which a substance may have the desired or toxic pharmacological effect depending on the stereoisomeric form. It is noteworthy that thalidomide was effective in the treatment of insomnia in leprosy patients, also showing anti-inflammatory, immunomodulatory and angiogenic activities [11].

After the pharmacophore and the group of the molecule responsible for the teratogenic effect were discovered [12], it became possible to design thalidomide-based hybrids that do not present any teratogenic effect. Inspired by thalidomide 4 and its anti-inflammatory and immunomodulatory activities, Lima et al. synthesized a series of hybrid substances containing N-substituted phthalimide groups and aryl sulfonamides 5 which were specially designed to selectively inhibit phosphodiesterase, an enzyme with an important role in the asthma process

(Scheme 2) [13]. Remarkably, the synthesized hybrids demonstrated considerable anti-asthmatic activity, with LASSBio-468 being the most active in the series, inhibiting lipolysaccharide (LPS)-induced neutrophil infiltration with an ED50 of 2.5 $mg.kg^{-1}$.

Another valuable example of the successful use of molecular hybridization is in anti-cancer therapy. Undoubtedly, cancer is to date the most lethal disease and one of the greatest evils of this century. Given the complexity of the various types of tumors, therapeutic combination is often ineffective due to pharmacokinetic (solubility, absorption, distribution and bioavailability) and pharmacodynamic (drug interactions or other than those of interest) factors. In this scenario, molecular hybridization can act by enhancing anticancer therapies. Estramustine, for instance, is an antineoplastic agent derived from the hybridization of estradiol with nornitrogen mustard (a mustard gas analogue), approved by the FDA in 1981 for the treatment of prostate cancer (Scheme 3) [14, 15]. Administered in the form of estramustine phosphate, it is readily dephosphorylated during absorption, releasing estramustine and, in spite of being derived from two alkylating agents, has its pharmacological action based on binding to microtubule-associated proteins (MAPs) and tubulin. As a result, there is a dynamic inhibition of microtubules and consequent arrest of cellular anaphase. Currently, this hybrid drug has been evaluated as a chemotherapeutic treatment for breast cancer, with ongoing advanced phase II tests for the evaluation of its effects on women with breast cancer.

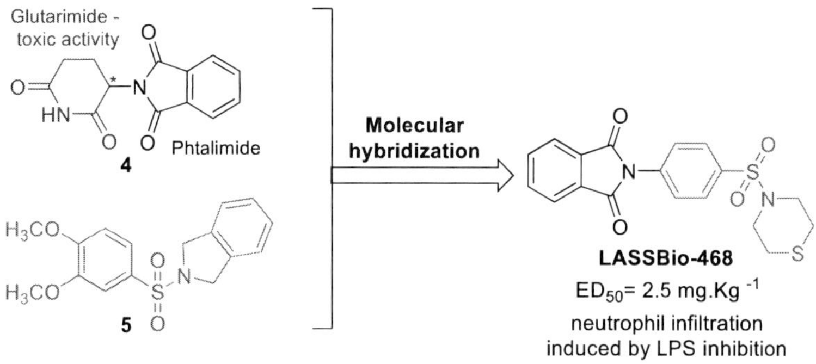

Scheme 2. Hybrid thalidomide-arylsulfonamide compound with anti-asthma activity.

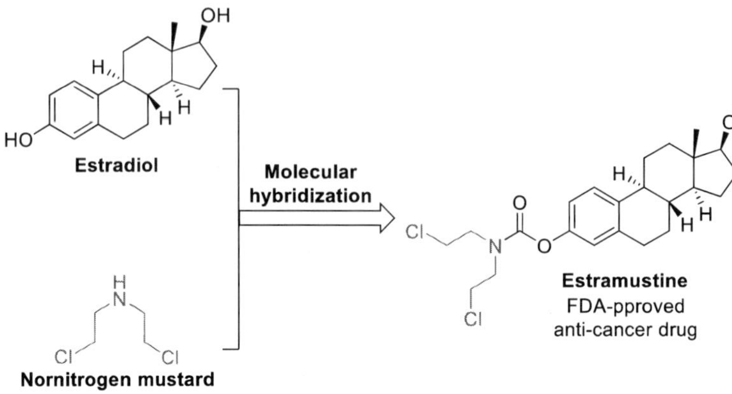

Scheme 3. Anti-cancer hybrids of estradiol and nornitrogen mustard.

An important example in this field is that of tacrine, which was the first drug approved by the Food and Drug Administration (FDA) in the United States for the treatment of Alzheimer's disease, acting as an inhibitor of the acetylcholinesterase (AChEI) enzyme. On the other hand, this compound has limited use due to its high hepatotoxicity [16, 17]. Cases like this are not uncommon, which turns therapies based on the "one molecule, one target" concept often difficult, given the complexities involved in such diseases. In this sense, molecular hybridization has proven to be an important ally in the fight against nervous system diseases such as Alzheimer's through the development of multifunctional hybrid drugs, many of which having tacrine as their core [17]. Nepovimova et al., for example, prepared hybrid molecules containing a tacrine derivative and 1,4-naphthoquinones (e.g., **9** and **10**) in an effort to combine tacrine's acetylcholinesterase inhibitory properties with the quinone's β-amyloid aggregation inhibiting activity (Scheme 4) [18]. As a result, a series of hybrid molecules was synthetized and evaluated, displaying IC_{50} values at nanomolar concentrations, with lower β-amylase aggregation inhibition rates than the standard drug. Compound **9**, specifically, had an IC_{50} value of 1.93 nM with an aggregation inhibition rate of 22.6%, while for **10** the IC_{50} value was 0.72 nM and the aggregation inhibition rate, 37.5% [18].

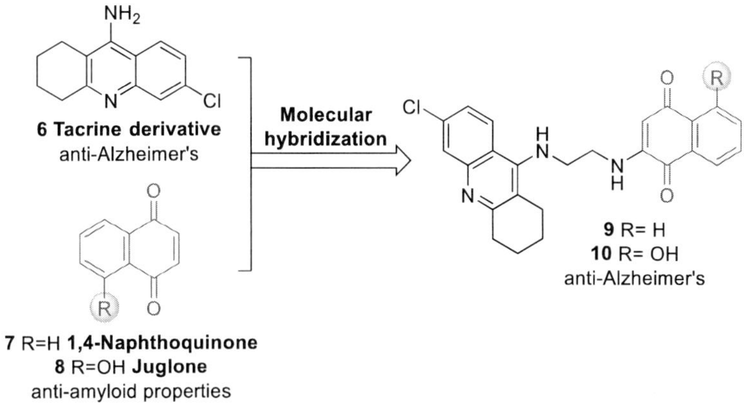

Scheme 4. Anti-Alzheimer hybrids of tacrine and naphthoquinones.

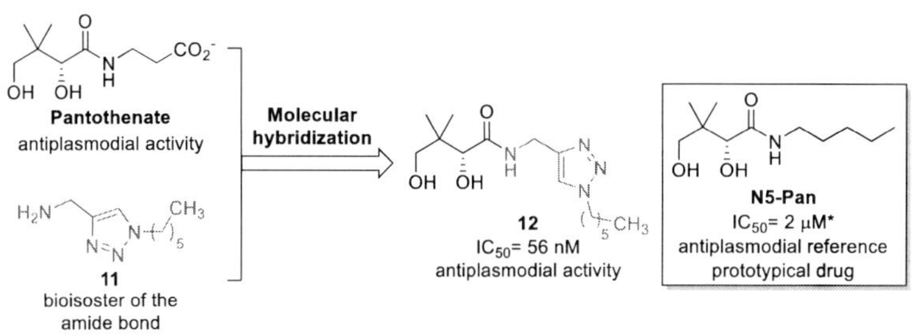

Scheme 5. Antispasmodic active pantothenamides-triazole hybrid. *in the absence of pantetheinase.

The combination of molecular hybridization with bioisosterism techniques has also been highlighted recently. Howieson et al., for instance, described the development of new antiplasmoidal compounds based on pantothenamide analog hybrids linked to 1-H-1,2,3-triazoles [19]. Pantothenate, also known as vitamin B5, is a biocompound required for the development of *Plasmodium falciparum* at its asexual stage in red blood cells, without which the parasite cannot survive [20, 21, 22]. In that way, pantothenate analogues have been reported to inhibit the growth of *P. falciparum,* but such substances were easily degraded by human plasma pantetheinases, inactivating the prototype drugs [23, 24, 25]. In order to overcome this limitation, the insertion of an amide bioisoster resistant to enzymatic hydrolysis into the structure was conducted. The most suitable group for this purpose was found to be the 1,2,3-triazole moiety. Following this concept, a series of substances was synthesized, with adduct **12** showing the best results in terms of IC50 (56 nM, Scheme 5). The activity of this compound was considerably higher than that of the reference prototype **N5-Pan.**

In summary, the aforementioned examples showcase the importance of molecular hybridization, with this technique being currently the cornerstone of medicinal chemistry in the search for new and more efficient drugs. Taking into consideration their growing importance in recent years, this chapter will focus on two classes of substances: 1,2,3-triazoles and naphthoquinones. Such scaffolds have been increasingly highlighted, mostly due to their pronounced biological activity and potential applications as new drugs. In this sense, this chapter aims to answer the following question: is it possible to combine 1,2,3-triazoles and naphthoquinones to produce new biologically active hybrids?

1,2,3-TRIAZOLES AS PHARMACOPHORIC FRAGMENTS

Five-membered heterocyclic rings containing one or more nitrogen atoms belong to the class of substances generically referred to as azoles. Even compounds with additional atoms other than nitrogen, such as oxygen or sulfur, are classified as azoles, being called oxazole and thiazole, respectively. Triazoles, in particular 1,2,3- triazoles, are the most studied azoles [26, 27, 28]. Importantly, in the disubstituted 1*H*-1,2,3-triazoles, two isomers are possible to occur: the 1,4- and 1,5-disubstituted (Figure 2). Considering that the preparation of 1,4-disubstituted-1*H*-1,2,3-triazoles is relatively easier than that of 1,5-disubstituted-1*H*-1,2,3-triazoles, most studies involving the synthesis and biological activities reported in the literature involve the 1,4- rather than the 1,5-isomer [29, 30].

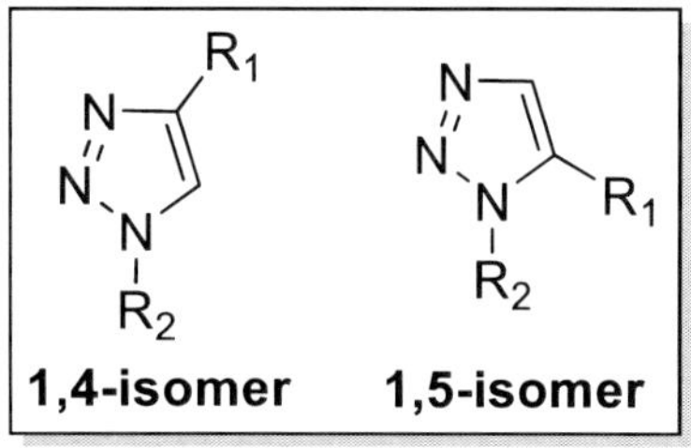

Figure 2. Structure of 1,4- and 1,5-disubstituted-1*H*-1,2,3-triazoles.

Over the last decades, several synthetic methods have been developed for the preparation of 1*H*-1,2,3-triazoles, starting from various reagents and different reaction conditions [4, 31]. Within these synthetic methodologies, perhaps the most important one is the thermal 1,3-dipolar cycloaddition between an alkyne and an aromatic or aliphatic azide, which is also known as the Huisgen azide-alkyne cycloaddition. Although very versatile, this reaction's main drawback is the lack of selectivity when non-symmetrical alkynes are used. In order to overcome this issue, Sharpless et al. were able to improve the regioselectivity towards the 1,4-disubstituted 1*H*-1,2,3-triazole isomer by introducing copper (I) as a catalyst [4, 32, 33]; in this case, the reaction proceeds with high yields, is mostly insensitive to the solvent, can be performed at room temperature, and is about 107 times faster than the uncatalyzed reaction. This synthetic method revolutionized the use of 1,2,3-triazoles in medicinal chemistry [34, 35] and has been reviewed by several authors, who report that it is a robust methodology that can be applied in combinatorial methods to generate large series of compounds [36, 37]. The mechanism of this reaction has also been thoroughly investigated and is well-established in the literature.

Interestingly, although the synthetic molecules that contain this unit show diverse biological activities, such heterocyclic moiety does not occur in nature. The synthetic studies and biological evaluation of these compounds indicate that their bioactivity is related to the functionalization of the groups around the ring [38, 39, 40, 41, 42]. Some authors have speculated that this range of biological activities has connection with the structure of the triazole ring, which has characteristics similar to those of a peptide bond, since they can accept hydrogen bonds and have a bonding distance of 3.8-3.9 Å, while for amides this value is 5.0-5.1 Å [43]. Additionally, the dipolar character of amides and the 1,2,3-triazole are 4.0 and 5.0 Debye, respectively [44, 45, 46]. The ability to participate in hydrogen bonds allied to its dipolar character favors the solubility of these heterocycles in the polar biological environment, as well as their interaction with biomolecular targets. All factors considered, several biologically active compounds with amide groups have already been replaced by triazoles with promising results [47].

Several compounds containing the 1,2,3-triazole scaffold are already in the pharmaceutical market. For instance, there are two β-lactam-1,2,3-triazole hybrids used as antibiotic, tazobactam and cefatrizine (Figure 3). Tazobactam came into clinical use in 1992 as an inhibitor of the bacterial action of β-lactamases, especially those belonging to the SHV-1 and TEM groups [48, 49]; it is commonly used as a sodium salt. The main use of this substance is as an adjunct to other antibiotics to make them less vulnerable to the antimicrobial resistance of bacteria [50]. Tazobactam is commonly found in combination with the antibiotic piperacillin, and is used in *Pseudomonas aeruginosa* infections. More recently [51], tazobactam has been released in combination with the cephalosporin antibiotic Ceftolozane [52] for the treatment of complicated urinary tract and intra-abdominal infections, as well as bacterial pneumonia. As for cefatrizine [53], it is a semisynthetic antibiotic available in several countries worldwide with broad-spectrum effectiveness against many positive and negative Gram microorganisms and infections at concentrations below 12.5 $\mu g.mL^{-1}$. Another FDA-approved 1,2,3-triazole that is in the pharmaceutical market since 2008 is rufinamide [54], a drug used for the treatment of various seizure disorders [55, 56, 57]. There is evidence that rufanamide acts by modulating sodium channels which are used as target for the development of antiepileptic drugs.

Besides some drugs that contain in the 1,2,3-triazole ring in their structure that have already reached the pharmaceutical market, there are also others in very advanced phases in their corresponding scientific studies. The human immunodeficiency virus (HIV-1) reverse transcriptase enzyme (RT) is a target that has been explored in the search for AIDS therapy drugs and, therefore, a large number of RT inhibitors, including the so-called non-RT inhibitors nucleosides (NNRTIs) have been reported. De Clerq et al., for instance, developed a series of thymine-derived compounds called TSAO that have demonstrated high RT inhibition. 1,2,3-Triazole analogues of TSAO (**13a-c**) have also showed significant activity (Figure 4) [58, 59, 60, 61, 62].

Carboxyamidotriazole, also known as CAI [63, 64], is an orally-active agent with potential antineoplastic activity that inhibits calcium channels, resulting in their disruption [65]. Also, the blockage of the calcium channel entry may inhibit cell proliferation, and consequently tumor cell growth, invasion and metastasis [66]. BRL-42715 has been extensively studied as an inhibitor of most bacterial β-lactamase enzymes, including cephalosporinases and penicillinases class I. The triazolic ring in this derivative confers to it a significant synergistic effect with Amoxylin in vivo and in vitro, comparable to clavulinic acid and sulbactam, known inhibitors of β-lactamase enzymes [67, 68, 69].

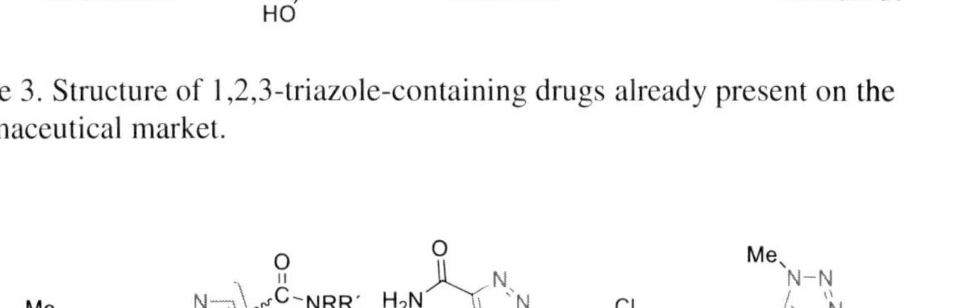

Figure 3. Structure of 1,2,3-triazole-containing drugs already present on the pharmaceutical market.

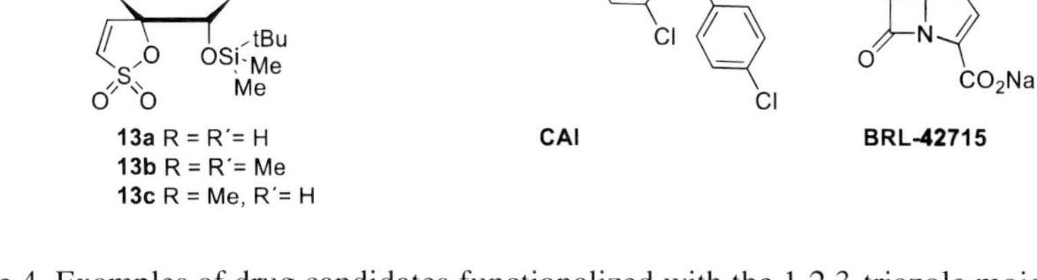

Figure 4. Examples of drug candidates functionalized with the 1,2,3-triazole moiety.

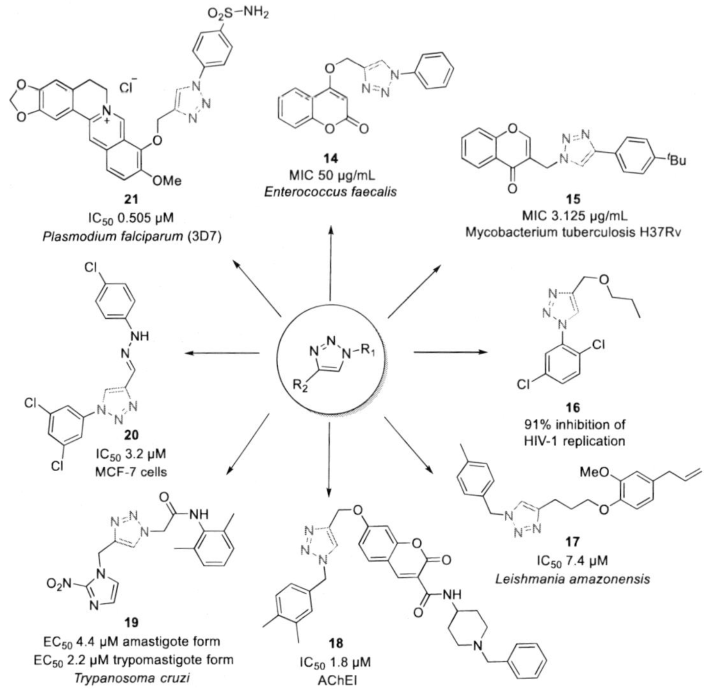

Figure 5. Selected examples of bioactive 1,2,3-triazole derivatives.

As the 1,2,3-triazole ring has shown to improve the activity of many compounds against various biological targets, and considering that several synthetic methods towards such scaffold are available, this heterocyclic ring has been widely explored in the development of new hybrid compounds. Some selected biological activities of 1,2,3-triazole-containing molecules are highlighted as, namely, antimalarial [70, 71, 72, 73, 74, 75, 76, 77, 78, 79, 80, 81, 82, 83, 84], antibacterial [85, 86, 87, 88, 89, 90], antifungal [91, 92, 93, 94, 95], antitubercular [15, 96, 97, 98, 99, 100, 101, 102, 103, 104, 105, 106, 107], antiviral [108, 109, 110, 111, 112, 113, 114, 115, 116], antileishmanial [117, 118, 119, 120, 121, 122, 123, 124], cholera toxin inhibition [125], cytotoxicity against cancer cells [126, 127, 128, 129, 130, 131, 132, 133, 134], anti-inflammatory [135, 136, 137, 138, 139, 140, 141, 142, 143], antiplatelet [144, 145, 146], trypanocidal [147, 148, 149, 150, 151, 152, 153], anti-allergic [154, 155, 156, 157, 158, 159], and anti-snake venon [160, 161] agents, among many others [162, 163,

164, 165, 166, 167, 168, 169]. Figure 5 shows the structure of some promising substances that were found in these studies.

Besides 1,2,3-triazoles, it is important to highlight that several drugs containing the 1,2,4-triazole moiety were developed and have also reached the pharmaceutical market (Figure 6) [170, 171].

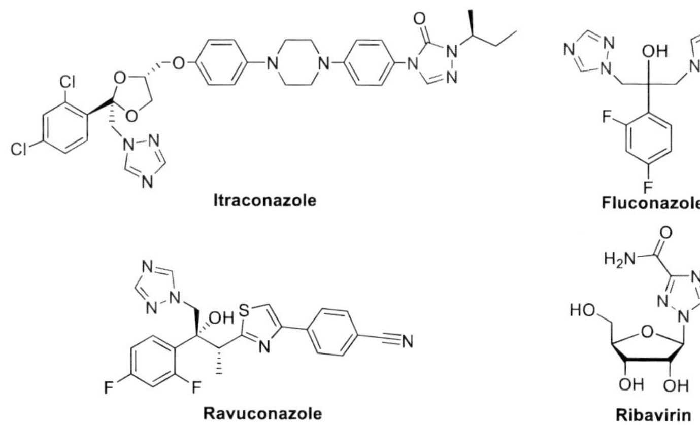

Figure 6. Examples of commercial drugs functionalized with 1,2,4-triazoles.

Finally, it is important mentioning that due to properties like resistance to metabolic degradation, stability to acidic/basic hydrolysis and reducing/oxidizing conditions, a consequence of their aromatic character [172, 173, 174, 175, 176], the 1,2,3-triazole ring has also been employed as a linker unit that can connect two pharmacophores to form innovative bifunctional drugs. Additionally, they have found applications in groundbreaking fields such as organocatalysis [177, 178, 179, 180, 181] and ionic liquids [182, 183, 184, 185].

THE NAPHTHOQUINONES

The naphthoquinones are an important class of compounds widely reported in the literature as having several biological activities. This class is divided into three main groups according to their different aromatic systems: benzoquinones, anthraquinones and naphthoquinones. The naphthoquinones, in specific, are derived from naphthalene and can be divided into two isomeric classes: 1,2- (*ortho*, 22, Figure 7) and 1,4-naphthoquinones (*para*, 23, Figure 7). These structural features leads to different chemical and physicochemical characteristics, as well as biological activities.

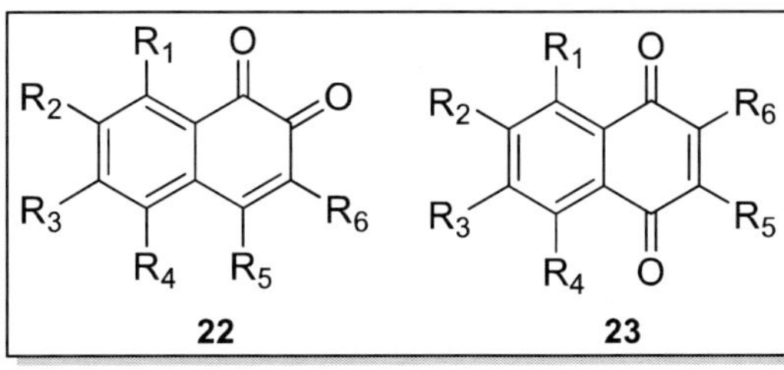

Figure 7. General structure of 1,2- and 1,4-naphthoquinones.

Naphthoquinones are secondary metabolites found in plants, animals, microorganisms, and marine organisms. Their applications vary from dyes, traditional phytomedicine and medicinal chemistry. A huge number of these synthetic and naturally occurring compounds, derivatives and analogues, have been synthesized and screened for their biological properties. 1,4-naphthoquinones, in specific, have been widely exploited due to their broad spectrum of biological activities, namely as antimicrobial [186], antitumor [187], antimalarial [188], antifungal [189] and antiviral agents [190], larvicides and molluscicides [191]. Their mechanisms of action depend on their structural features, and have been recently highlighted for cancer cells [192, 193].

There are several natural and synthetic naphthoquinone-based drugs of clinical utility that are in the market (Figure 8). Lapachol is a natural naphthoquinone present in plants of the genus *Tabebuia* that has been used in the past for the treatment of Walker 256 carcinoma [194] and Yoshida sarcoma [195]. Another important structural analog of lapachol is isolapachol. The toxicity of soluble isolapachol derivatives (potassium salt) was evaluated in different stages of the life cycle of *Schistosoma mansoni*, and the compound exhibited significant activity against adult forms, cercariae and eggs [196]. In addition, this molecule was assayed against metacyclic promastigotes of *L. amazonensis* and *L. brasiliensis*, two different species of *Leishmania* associated with tegumentar leishmaniasis, showing considerable activity [197]. Atovaquone **8** is a 2-hydroxy-1,4-naphthoquinone derivative currently in use as a therapeutic drug in the treatment or prevention of mild cases of infection by *Plasmodium vivae* (in combination with proguanil) [198, 199, 200, 201], although there have been reported cases of resistance [202]. Recently, a secondary used was devised for this drug for pneumocystosis, toxoplasmosis and babesiosis (usually in combination with azithromycin). Buparvaquone and parvaquone are in use for the treatment of bovine theileriosis and bovine East Coast fever protozoa (Figure 8) [203]. It is noteworthy that atovaquone, buparvaquone and parvaquone were synthesized with basis in the structure of lapachol, all having the same core but different alkyl side chains. Also, these drugs have similar pharmacology [204]. Many other natural naphthoquinones have already been isolated and evaluated for their potential biological activities. Two interesting examples are salinisporamycin and shikonin. Salinisporamycin is a rifamycin antibiotic that has been isolated from cultures of marine actinomycete *Salinispora arenicora*; it inhibited the growth of human lung adenocarcinoma cell A549 with an IC_{50} of 3 $\mu g.mL^{-1}$, also having shown antimicrobial activity. Shikonin is a natural 1,4-naphthoquinone extracted from the roots of *Lithospermum erythrorhizon* that inhibits cancer cell glycolysis by targeting tumor pyruvate kinase-M2 and the migration of ovarian cancer cells, inducing their apoptosis by impairing the phosphorylation of two kinases [205, 206]. Streptonigrin is a natural naphthoquinone isolated from *Streptomyces flocculusin* that has attracted considerable attention not only due to its activity against several cancer cell lines, but also its antibiotic activity [207]; it reached phase II clinical trials, but those were abandoned due to the high levels of toxicity and side effects [208]. The 1,4-naphthoquinone ring is also present in many natural antibiotics, e.g., rifamycin S [209], kinamycin C [210] and streptovaricin U [211] (Figure 8).

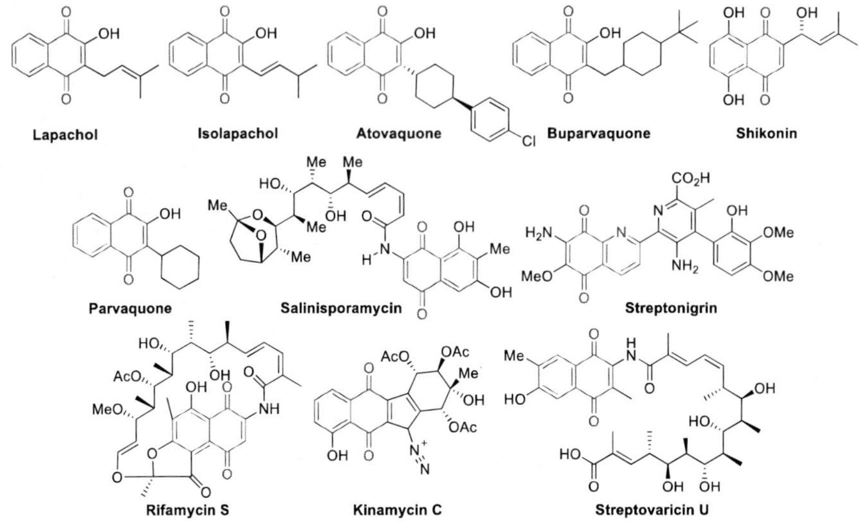

Figure 8. Structure of a few important biologically active naphthoquinone-based drugs and candidates.

1,2-Naphthoquinones also have important reported biological activities owing to their high affinity to form covalent bonds with proteins and DNA, interfering in the cellular cycle. In addition, it functions as a pro-oxidant, forming reactive oxygen species (ROS). This nucleus is present in several natural products that exhibit pronounced biological activities (Figure 9) [158]. Saprorthoquinone is a diterpene isolated from *Salvia hypargeia*, and its cytotoxic activity in human colon cancer has been known for over two decades [212]. Tanshinone I is a phenanthrenequinone derivative isolated from *Salvia miltiorrhiza Bunge*, having anticancer activity against human gastric cancer cells [213]. Miltirone is used in traditional Chinese medicine, and has a myriad of biological activities such as anti-inflammatory, antioxidant and pain-relieving, as well as *in vitro* antineoplastic activity against various types of cancer [214]. β-Lapachone, which is currently under a phase 3 clinical trial, is a derivative obtained from lapachol (Figure 8) and has been known for its antiparasitic, anti-inflammatory, antibacterial, and antitumor activities [215].

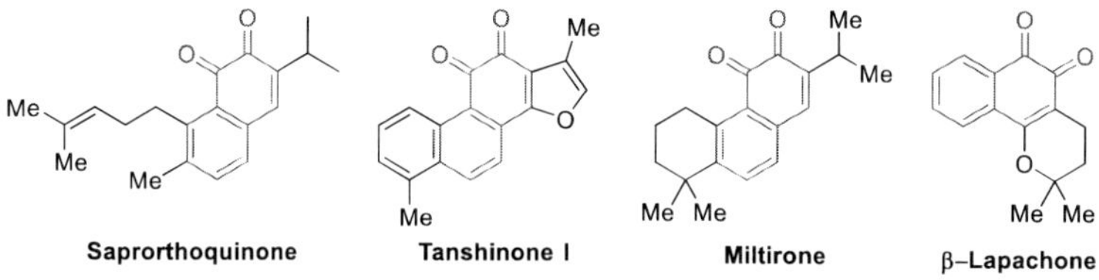

Figure 9. Structure of important 1,2-naphthoquinone-containing compounds.

Besides their biological importance, naphthoquinones are widely used as starting materials in organic synthesis. There are many methods for the preparation of such compounds, the most convenient starting materials being naphthol derivatives, which are easily obtained from naphthalene. 1,2-Naphthoquinones are usually prepared from the oxidation of 1,2-aminonaphthol hydrochloride **24** in acidic solution (Scheme 6, route a). However, the preparation of this intermediate in good yields and adequate purity using this route can be challenging [216]. Recently, Sim and co-workers have described a method for obtaining 1,2-naphthoquinones from 2-naphthol **25** through a *N*-bromosuccinimide (NBS)-mediated sequential bromination reaction followed by oxidation in the presence of piperidine (Scheme 6, route b). This simple one-pot procedure is very promising, since it does not require the use of toxic metals or expensive materials [217]. Another direct and efficient methodology towards 1,2-naphthoquinones using **25** was described by Shukla et al., in which a copper-catalyst is used in the presence of TEMPO (Scheme 6, route c) [218].

As for 1,4-naphtoquinone, the most convenient method for its preparation is the oxidation of 4-amino-1-naphthol hydrochloride **26** (Scheme 7, route a) [219], but it can also be prepared via the oxidation of naphthalene **27** (Scheme 7, route b) [220], and 2-naphthol **25** (Scheme 7, route c) [221]. Recently, Gogin et al. described a new method for the synthesis of 1,4-naphthoquinone through a one-pot reaction between hydroquinone **28** and different 1,3-dienes in aqueous solutions using Mo-V-P heteropolyacids HPA-x (x = 4, 7, 10) as bifunctional catalysts (Scheme 7, route d) [222].

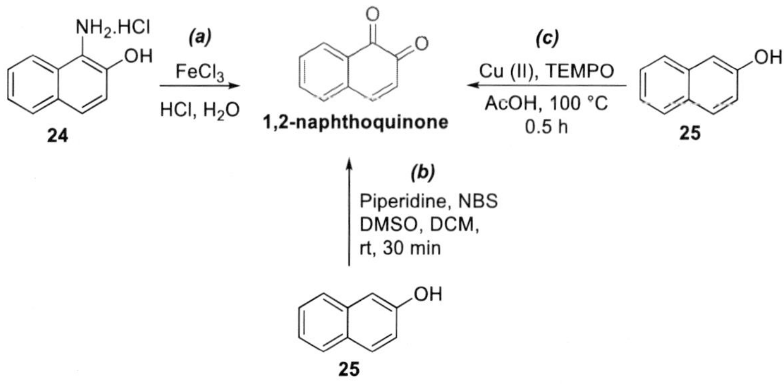

Scheme 6. Selected methods for the preparation of 1,2-naphthoquinones.

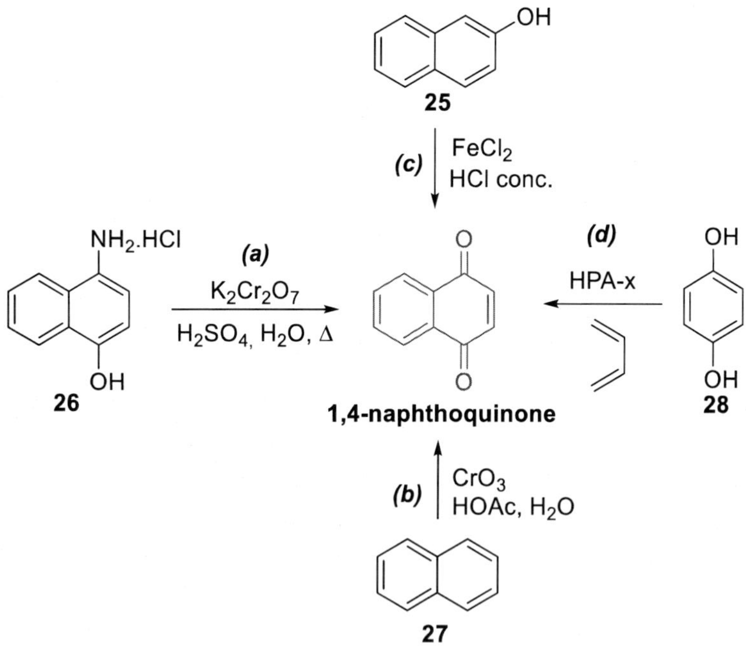

Scheme 7. Selected methods for the preparation of 1,4-naphthoquinones.

The great synthetic versatility of naphthoquinones arises from the presence of several functional groups in their structure, namely two carbonyls, a C-C double bond and an aromatic ring. The carbonyl groups can serve as anchors for vicinal reactions such as iodination [223], thiolation [224] and selenilation [225] that take place at the double bond and the aromatic ring. Because of the presence of carbonyls, the benzene ring is deactivated towards aromatic electrophilic substitution reactions, but the vicinal position to the carbonyl groups can be reacted with electrophiles under transition metal catalysis. Indeed, several reactions have been explored at the C-C double bond, such as Michael addition [226], epoxidation [227], 1,3-dipolar cycloaddition [228], Diels-Alder [229], α-arylation [230], and alkylation [231].

Recently, two substituted β-2,3-dihydrofuranaphthoquinones (**62** and **63**) showed antifungal activity against *Sporothrix brasiliensis* and *Sporothrix schenckii*, the main etiological agents for sporotrichosis in Brazil (Figure 10). The dihydro-furanonaphthoquinones showed good antifungal activity and stability, with MIC values of 2–16 μM. Additionally, these compounds induced yeast-hyphae conversion and alterations in the hyphae and conidia structures in fungi. The performed tests indicated that such molecules are potential candidates for the development of new antifungal agents for the treatment of sporotrichosis [232].

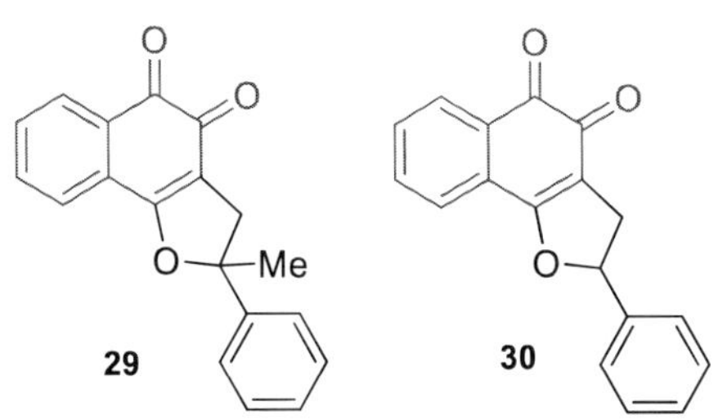

Figure 10. Structure of β-2,3-dihydrofuranaphthoquinones with promising antifungal activity.

Some patents published in 2018-2019 involving pharmaceutical compositions of naphthoquinones highlighted different molecules, namely a peracetylated fucoside, called flau-a [2-(2,3,4-tri-o-metil-6-deoxy-β-L-galactopyranosyloxy)-1,4-naphthoquinone] (**31**, Figure 11), which showed selective leishmanicidal activity against promastigotes and amastigotes forms of leishmania [233], the naphthoquinone β-D-glucoside [3-(3-methylbutene-3-yl)-6,7-dimethoxy-naphthoquinone-2-O-β-D-glucoside] (**32**, Figure 11), which is as a potential candidate for improving the senile dementia (CN110025624) [234], and 6,11-dimethoxy-15,16-dimethyl furan-naphthodioxene-1,2,14-trione (**33**, Figure 11), in preparations of blood sugar reducing products (CN108503653) [235].

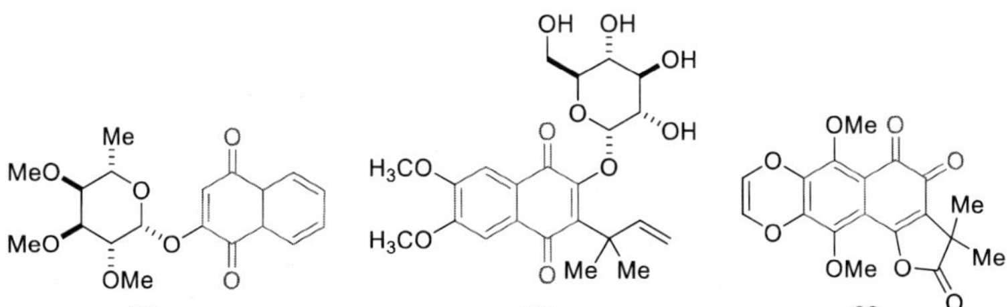

Figure 11. Structure of some drugs and candidates based on the naphthoquinone nucleus.

Naphthoquinone-1,2,3-Triazole Hybrids

The incorporation of a triazole ring into the naphthoquinone scaffold was first achieved by Ferreira and et al, which obtained a series of new hybrid compounds and assayed them against the infective bloodstream trypomastigote form of *T. cruzi* [236]. Remarkably, all the hybrid derivatives were more active than the corresponding quinones, with IC_{50} values in the range of 17 to 57 μM; the two most active compounds were substituted triazoles **36a** and **b** (Scheme 8). This series of triazole-naphthoquinones have emerged as interesting new lead compounds in drug development for Chagas disease. It is worthy to note that these β-furanaphtoquinone-1,2,3-triazoles also have potential clinical utility in the treatment of human cancer. Indeed, they had their cytotoxic activities evaluated against six neoplasic cancer cell lines: SF-295 (central nervous system), HCT-8 (colon), MDAMB-435 (melanoma), HL-60 (leukaemia), PC-3 (prostate) and B-16 (murine melanoma) and against one normal cell, the murine fibroblast L-929, and the same triazoles **36a** and **b** showed high and selective activity against melanoma cells MDAMB-435 [237]. A similar molecular hybridization approach was extended to the synthesis of α-furanonaphthoquinone linked to 1,2,3-triazoles (**38a-b**). Such molecules were also tested against Leishmania parasites and the authors observed that the insertion of a 1,2,3-1*H*-triazole moiety into the α-furanonaphthoquinone enhanced its leishmanicide activity when compared to nor-α-lapachone (Scheme 8) [238].

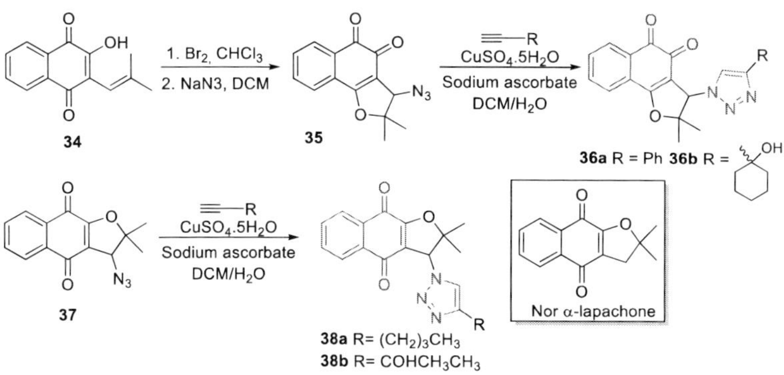

Scheme 8. Hybrids of furan and pyranonaphthoquinones with 1,2,3-1*H*-triazoles.

Several ruthenium catalysts are known to alter the regioselectivity of the Huisgen 1,3-cycloaddition, producing the 1,5-substituted product rather than 1,4-, which is obtained in the presence of copper(I) catalysts. With these concepts in mind, Da Silva Junior et al. synthesized a series of 1,5-disubstituted-1,2,3-triazoles linked to β-furan- and β-pyranona-phthoquinones **39** and **41** (Scheme 9) [239]. Besides the evaluation against *Trypanossoma cruzi*, six cancer cell lines were also evaluated: PBMC (lung), PC3 (prostate), HCT-116 and HCT-8 (colon), MDA-MB-435 (breast) and SF-295 central nervous system. Two compounds in the series were the most active against all tested cancer cell lines, and both presented activity equivalent to that of benznidazole.

Da Silva Junior et al. also reported that the introduction of the 1,2,3-triazole ring directly into the 1,4-naphthoquinone scaffold, which led to hybrids **76,** enhanced the trypanocide activity when compared to the corresponding 1,4-naphthoquinones (Scheme 10a) [240]. Employing the same methodology, Faria et al. used juglone (**42**, R = OH, Scheme 10) as starting material, and new 1,2,3-triazole-naphthoquinones hybrids were synthesized and tested on their inhibition of the P_2X_7 receptor, which is involved in inflammatory processes [241]; two compounds (**43a** and **b)** presented high anti-inflammatory activity with low toxicity (IC_{50} 0.23 and 0.103 μM, respectively), being better than the standard A740003 (Scheme 10b).

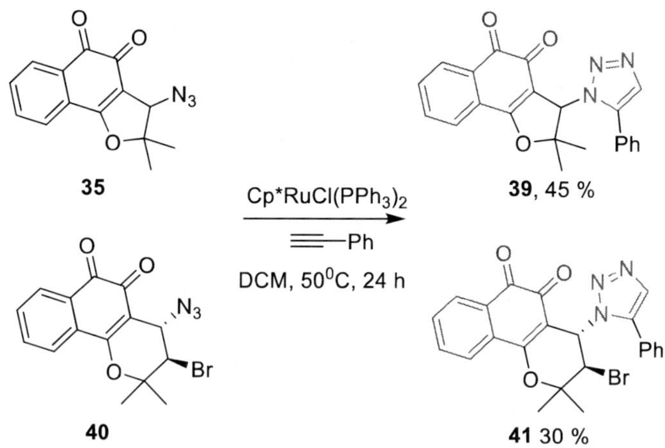

Scheme 9. Synthesis of pyran and furanonaphthoquinone-1,5-disubstituted-1,2,3-triazoles hybrids.

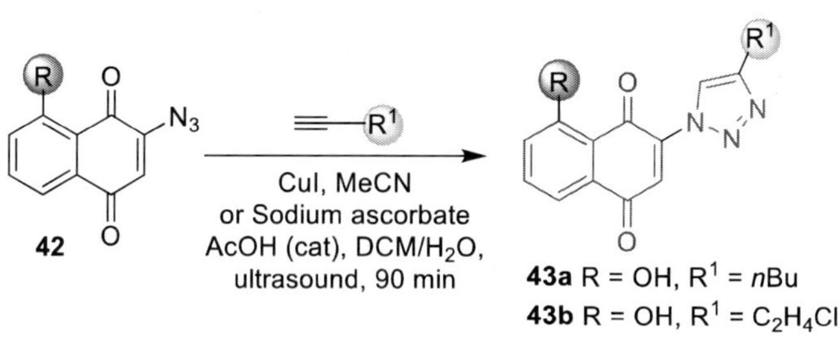

Scheme 10. Hybrids of 1,4-naphthoquinones directly linked to 1,2,3-triazoles.

The introduction of one or two alkyl- or aryl-amino substituents into the naphthoquinone nucleus generates new families of compounds with amplified biological activity (Figure 12). Indeed, several studies indicated that amino-1,4-naphthoquinone derivatives display increased biological activities when compared to the original naphthoquinones.

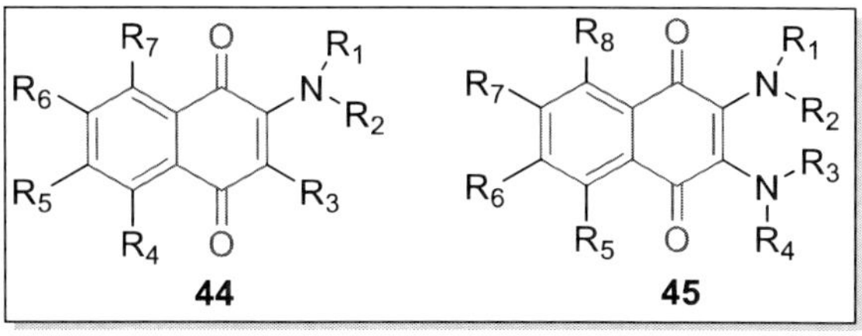

Figure 12. General structure of amino-1,4-naphthoquinone derivatives.

Typically, aminonaphthoquinones such as **44** and **45** can be further converted into other versatile synthetic intermediates. Da Silva et al., for instance, synthesized propargylated intermediate **46** and used it to generate a series of *para*- and *ortho*-naphthoquinones coupled to 1,2,3-triazoles (Scheme 11). The compounds were evaluated against the infective bloodstream form of *Trypanosoma cruzi*, and three of them were more active than standard benznidazole (**50b** and **c**, IC_{50} 6.8 and 8.2 μM, respectively). The same group has also synthesized a 1,2,3-triazole scaffold linked to two aminonaphthoquinones **48,** and to a 1,4-naftoquinone and a carbohydrate **49**. Such compounds had their cytotoxic activities evaluated against thirteen neoplasic cancer cell lines - OVCAR-8 (ovary), SF-295 (central nervous system), HCT-8 and HCT-16 (colon), DU-145 (prostate), MDAMB-435, MCF-7, HS5787, MXL (breast), HL-60 (leukemia), PC-3 (prostate), B-16 (murine melanoma), PBMC (lymphoma), V79 (lung) and against one normal cell, the murine fibroblast L-929, having showed promising results [242].

Recently, Khoshneviszadeh et al. designed and developed a series of new 1,4-naphthoquinone-1,2,3-triazole hybrids **47** by reacting amino-1,4-naphthoquinone **46** with azidomethyl benzene. The compounds were evaluated against three cancer cell lines (MCF-7, HT-29 and MOLT-4), and all compounds showed good cytotoxic activity. Compound **47a** stood out for its excellent activity and selectivity against HT-29 cells (IC_{50} 6.8 μM), being more potent than the cisplatin standard [243].

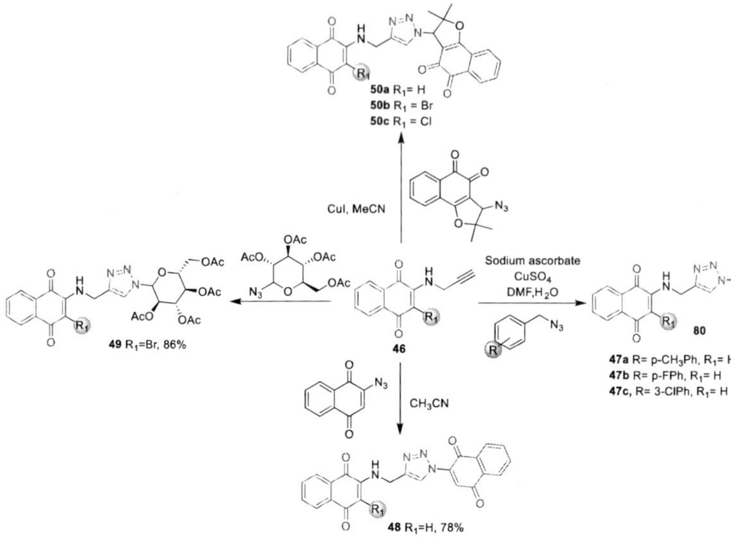

Scheme 11. Amino-1,4-naphthoquinones as starting materials for the synthesis of new 1,2,3-triazole hybrids.

Another alternative for the synthesis of new hybrid molecules is the *N*-alkylation of primary amino-1,4-naphthoquinones. Da Silva Junior et al. used this strategy to obtain propargyl-functionalized secondary aminonaphthoquinones **52**, which were further converted into 1,2,3-triazoles **53** via a click reaction with tosyl azide. The triazoles **53** were evaluated against eight types of cancer cell lines, and some of them exhibited potent cytotoxicity with IC_{50} values below 1.0 μM (Scheme 12) [244].

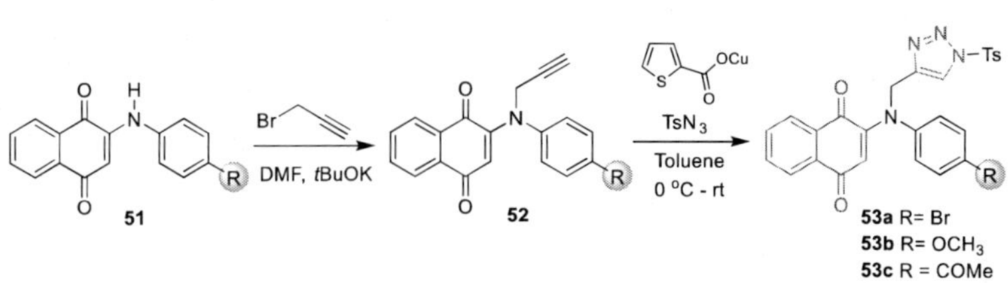

Scheme 12. Synthesis of triazole-amino-1,4-naphthoquinones hybrids with activity against cancer cells.

Besides *N*-propargylated aminonaphthoquinones, their *O*-propargylated counterparts have also been widely explored. Some groups have used this approach in order to obtain new 1,2,3-triazole-naphthoquinone hybrids with antimalarial activity against *P. falciparum*. In this case, both propargylated lapachol **54** [245] or lawsone **56** could be used in 1,3-dipolar cycloaddition reactions with different organic azides. Compounds **55a**, **b** and **c** proved to be the most active against the protozoan and the least cytotoxic.

In another study which used a similar methodology, Oramas-Royo et al. obtained several hybrid compounds which were evaluated against six bacterium species: *Streptococcus pneumoniae*, *Enterococcus faecalis*, *Escherichia coli*, *Klebsiella pneumoniae*, *Pseudomonas aeruginosa* and *Salmonella enterica*. Compound **57a** presented the highest activity against S. *pneumonia* (IC_{50} 39.33 μM) and *E. coli* (IC_{50} 22.57 μM), while **57b** was more active against *E. fecalis* (IC_{50} 39.23 μM). Among these hybrid compounds, **57c** exhibited the most potent activity against *P. falciparum* (Scheme 13) [246].

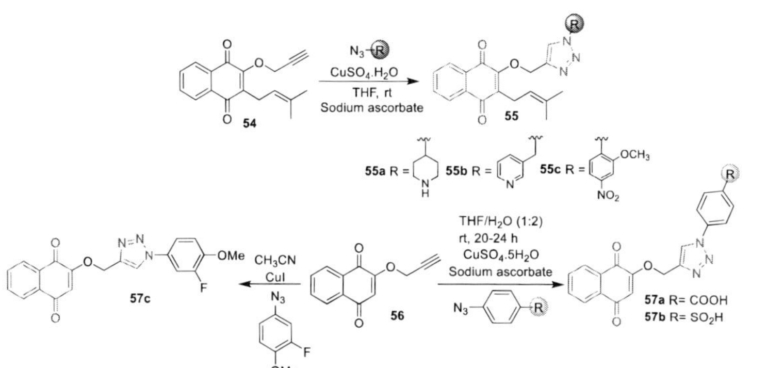

Scheme 13. 1,4-Naphthoquinones-1,2,3-1*H*-triazoles with antimalarial activity.

Other naphthoquinone derivatives have also been used for conjugation with 1,2,3-triazoles. Da Silva Junior et al., for example, explored menadione **58** to obtain 1,2,3-triazole derivatives of 1,4-naphthoquinones [247]. In order to accomplish that, menadione **58** was oxidized with hydrogen peroxide in methanol to give 2-hydroxy-3-methyl-1,4-

naphthoquinone, which was subsequently methylated, leading to intermediate **59** with an overall yield of 48% (Scheme 14). Next, an azide group was inserted into the structure after bromination with NBS followed by displacement with sodium azide, leading to compound **60**. Then, naphthoquinone azide **60** reacted with different substituted phenylacetylenes in the presence of a copper (I) catalyst, affording products **61** in good to moderate yields. All compounds were evaluated against 5 tumor cell lines, HL-60 (leukemia), HCT-116 (colon), SF295 (nervous system), NCIH-460 (lung) and PC-3 (prostate), and two of them (**61a** and **61b**) demonstrated to be selective for HL-60 cells with low toxicity for normal cells L929 (Scheme 14).

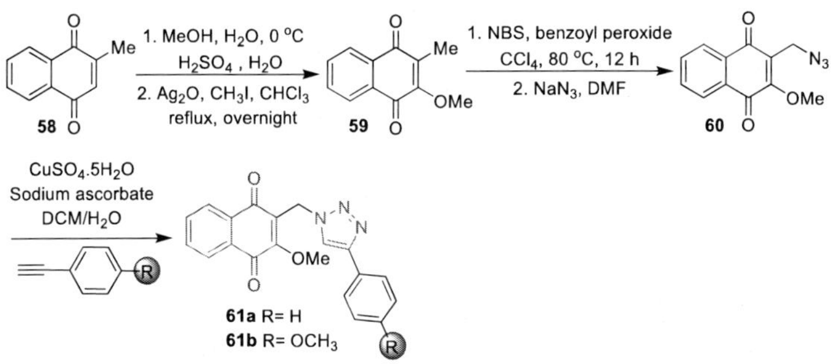

Scheme 14. Synthesis of methoxy menadione-1,2,3-triazole hybrids.

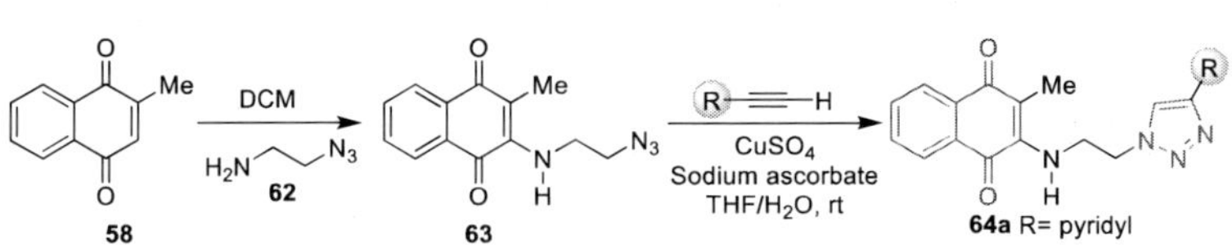

Scheme 15. Synthesis of 1,2,3-triazole-menadione hybrids.

Menadione was also used by Mallavadhani et al., which used this molecule as a Michael acceptor in the presence 2-azidoethylamine **62** to give product **63** (Scheme 15) [248]. 1,4-Naphthoquinone azide **63** was reacted with different terminal alkynes to form hybrid compounds **64**, which were tested against cancer cell lines A549 (lung), DU-145 (prostate), Hela (cervical) MCF-7 (breast) and B-16 (melanoma).

Admirably, the hybrid triazole-naphthoquinone **64a** demonstrated to be more active than tamoxifen against all types of cancer cells.

Many studies have shown that organic compounds containing selenium have amplified biological activity. For example, a large variety of selenium derivatives have shown to have activity against a broad range of microorganisms, including bacteria and fungi [249, 250]. In this context, two important azido-naphthoquinone derivatives were prepared and underwent 1,3-dipolar cycloaddition with alkyne-selenium reagents (Scheme 16, **65** and **67**), leading to the formation of two types of new triazole-1,2-naphthoquinones hybrids (Scheme 16, **66** and **68,** respectively). These compounds were evaluated against some cancer cell lines - HL-60 (leukemia), HCT-116 (colon), SF295 (nervous system), PBMC, V79 or NCIH-460 (lung), OVCAR-8 (ovario), PC-3 (prostate), MDAMB-231 (breast) or PC-3 (prostate) SF295(nervous system) – and have showed promising results. Alternatively, an alkyne can be inserted into the 1,2-napththoquinone structure via an amine linker as in compound **69**, which can next react via click reaction with selenium-functionalized azide **70**, leading to the formation of product **71**. Compound **71** has displayed high biological activity against several cancer cell lines (Scheme 16) [251].

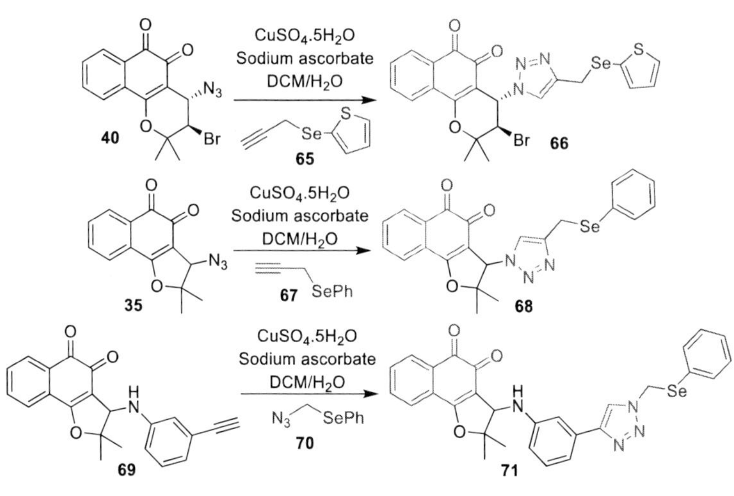

Scheme 16. Synthesis of selenium-containing lapachone-1,2,3-triazole hybrids.

Other hybridization approaches include electrocyclic processes such as the one disclosed by Da Silva and et al., [252] which employed this strategy to achieve triazole-naphthoquinones hybrids **73a** and **b** from 4-vinyl-1,2,3-triazole **72** and lawsone (Scheme 17). Despite its utility, this reaction is not highly regioselective, giving both the 1,4 and 1,2-naphthoquinonic hybrids. The compounds were evaluated against three human cancer cell lines - Caco-2 (colon), Calu-3 (lung) and MDA-MB231 (breast) - and the absence of toxic effects was observed in Vero cells (African green monkey kidney epithelial cells). Importantly, compound **73b** presented selective activity against Caco-2 cells with low toxicity against Vero cells.

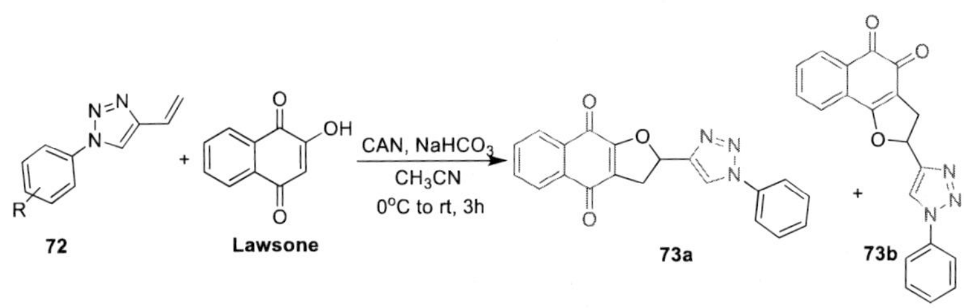

Scheme 17. Radical electrocyclic reaction for the synthesis of α-nor- and β-nor-lapachones linked to 1,2,3-triazoles.

In continuation of the efforts to identify potential anticancer agents with novel types of structures which are distinct from those of existing drugs, Da Silva et al. prepared different 1,2,3-triazole hybrids with analogs of nor-β-lapachone [253]. Starting from the C-3 alkylation of lawsone with allyl bromide in basic medium to give **74**, the later compound was transformed into the iodide derivative **75a** and subsequently to azide-1,2-naphthoquinone **75b**. Finally, the azide-naphthoquinone **75b** was submitted to a Cu(I)-catalyzed 1,3-dipolar cycloaddition with different commercial alkynes (Scheme 18). Two compounds stood out as the most active and selective (**76a** and **b)** against cancer cell MCF-7 (breast cancer) and HCT-111 (colon cancer).

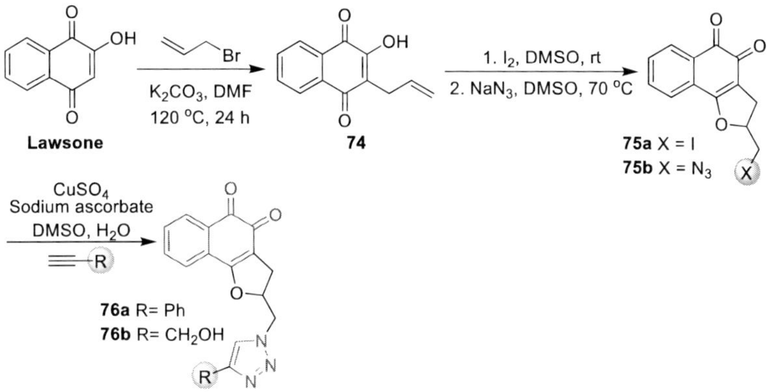

Scheme 18. Cytotoxic nor-β-lapachone-1,2,3-1*H*-triazoles hybrids.

Lawsone was used once again by Ferreira et al., which recently reported the preparation of *bis*-naphthoquinones from the reaction of lawsone with aldehydes in the presence of a catalytic amount of *p*-toluenesulfonic acid [254]. Da Silva et al. applied this approach for substituted 4-aldehyde-1,2,3-triazoles **77**, obtaining *bis*-naphthoquinone-1,2,3-triazoles **78**, which underwent a sulfuric acid-catalyzed intramolecular cyclization, giving products **79** (Scheme 19) [255]. Compounds **79a-d** showed promising anticancer activity against MCF-7 breast cancer line, with the hybrid **79a** being selected for further studies due to its capacity to produce ROS and reduce cell viability. The experiments proved that **79a** had anticancer activity, acting via the generation of ROS, which caused interference in the energy metabolism, cell viability and proliferation, in addition to cell membrane damage.

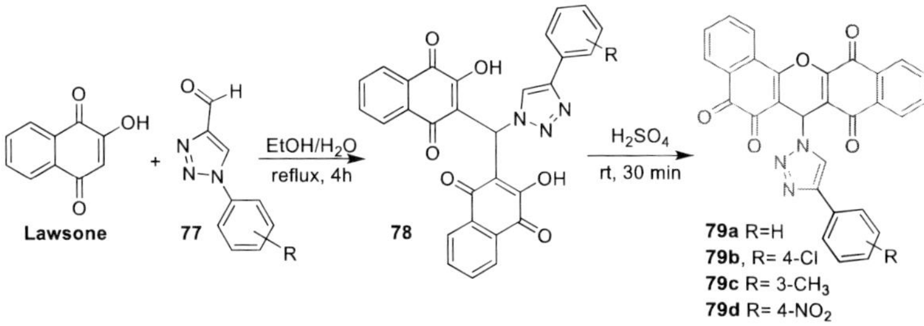

Scheme 19. Synthesis of *bis*-naphthoquinone-1,2,3-triazole hybrids.

The 1,2,3-triazole was also used as a linker between 4-quinolones and 1,4-naphthoquinones [256]. 4-Quinolones are a large group of molecules with antibiotic activity against Gram-negative and Gram-positive bacteria; many 4-quinolones are in the human and veterinary pharmaceutical market due to their broad spectrum of action. Many quinolones have been coupled with other nuclei in the search for new antimicrobial substances such as ciprofloxacin, a second-generation antibiotic. In this sense, the conjugation of the 4-oxoquinolinic nucleus with 1,4-naphthoquinones using 1,2,3-triazoles as linkers was performed, and the resulting compounds were evaluated against different cancer cell lines. For that, propargylated 1,4-naphthoquinone **80**, which was obtained from lapachol, was reacted with azido-oxoquinolinic ester **81,** producing the hybrid **82** (Scheme 20)**.** Such compound displayed strong and selective antitumor properties against the MCF-7, MDA-MB-231 and 4T1 (breast cancer) cancer cell lines. Additionally, further experiments have shown that four pathways are involved in the mechanism of action of this compound: glycolysis inhibition, decreased intracellular ATP levels, ROS induction and apoptosis triggering.

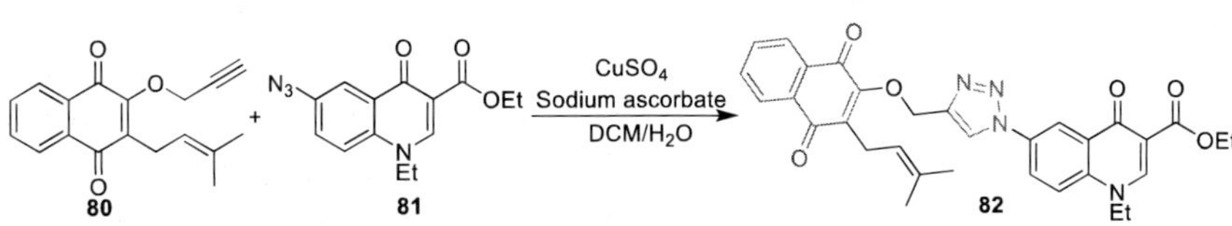

Scheme 20. Synthesis of a 1,2,3-triazole-linked 1,4-naphthoquinone-4-oxoquinoline hybrid.

Conclusion

In recent years, molecular hybridization has become an indispensable tool in the arsenal of medicinal chemists. As a consequence, the combination of several classes of molecules has been achieved, giving rise to more active and selective compounds.

Among the several classes of compounds that have been combined, the junction of the 1,2,3-triazole scaffold with 1,4-naphthoquinones has undoubtedly been highly fruitful. The natural occurrence of a variety of 1,4-naphthoquinones is a great advantage of working with this type of compound, as well as their versatile reactivity, a consequence of the diverse functional groups present in their structure. As a consequence, several new hybrid compounds were obtained in recent years, having showed promising activity not only against aggressive cancer cell lines, but also in the treatment of neglected diseases such as Chagas and Leishmaniosis.

On the other hand, a wide window of opportunities is still open in what regards the design and synthesis of new 1,2,3-triazole-naphthoquinone hybrids. Furthermore, the development of modern synthetic methodologies for the synthesis of such compounds has also become a fast-growing field. In our view, this field has already been blooming in recent years, but there is still a lot of room for growth.

REFERENCES

[1] Jacobus Ariëns Everhardus (1988). Stereochemical implications of hybrid and pseudo hybrid drugs. Part III. *Medicinal Research Review* 8(2): 309-320.

[2] Viegas-Junior Cláudio, Amanda Danuello, Vanderlan da Silva Bolzani, Eliezer J. Barreiro, Carlos Alberto Manssour Fraga (2007). Molecular hybridization: A useful tool in the design of new drug prototypes. *Current Medicinal Chemistry*, 14(17): 1829-1852.

[3] Ivasiv Viktoriya, Claudia Albertini, Ana E. Gonçalves, Michele Rossi, Maria L. Bolognesi (2019). Molecular Hybridization as a Tool for Designing Multitarget Drug Candidates for Complex Diseases. *Current Topics in Medicinal Chemistry*, 19: 1694-1711.

[4] Biot Christophe, Kelly Chibale (2006). Novel approaches to antimalarial drug discovery. *Infectious Disorders e Drug Targets*, 6: 173-204.

[5] Tietze Lutz F., Hubertus P. Bell, Chandrasekhar Srivari (2003). Natural product hybrids as new leads for drug discovery. *Angewandte Chemie International Edition*, 42: 3996-4028.

[6] Gediya Lalji K., Vincent C. O. Njar (2009). Promise and challenges in drug discovery and development of hybrid anticancer drugs. *Expert Opinion on Drug Discovery*, 4: 1099-1111.

[7] Morphy Richard, Corinne Kay, Zoran Rankovic (2004). From magic bullets to designed multiple ligands. *Drug Discovery Today*, 9: 641-651.

[8] Hulsman Niels, Jan P. Medema, Carina Bos, Aldo Jongejan, Rob Leurs, Martine J. Smit, I Wan J. de Esch, Dick Richel, Maikel Wijtmans (2007). Chemical insights in the concept of hybrid drugs: the antitumor effect of nitric oxide-donating aspirin involves a quinone methide but not nitric oxide nor aspirin. *Journal of Medicinal Chemistry*, 50: 2424-2431.

[9] Marco-Contelles José, Elena Soriano (2011). The medicinal chemistry of hybrid-based drugs targeting multiple sites of action. *Current Topics In Medicinal Chemistry*, 11: 2714-2715.

[10] Nepali Kunal, Sahil Sharma, Manmohan Sharma, Preet M. S. Bedi, Kanahya L. Dhar (2014). Rational approaches, design strategies, structure activity relationship and mechanistic insights for anticancer hybrids. *European Journal of Medicinal Chemistry*, 77: 422-487.

[11] Sheskin Jacob (1965). Thalidomide in the treatment of lepra reaction. *Clinical Pharmacology & Therapeutics*, 6: 303-306.

[12] Lima Lídia M., Carlos Alberto M. Fraga, Eliezer J. Barreiro (2001). O renascimento de um fármaco: Talidomida. *Quimica Nova*, 24: 683-688.

[13] Lima Lídia M., Paulo Castro, Alexandre L. Machado, Carlos Alberto M. Fraga, Claire Lugnier, Vera Lúcia G. Moraes, Eliezer J. Barreiro (2002). Synthesis and anti-Inflammatory activity of phthalimide derivatives, designed as new thalidomide analogues. *Bioorganic Medicinal Chemistry*, 10: 3067-3073.

[14] Hauser Allen R., Ruth Merryman (1984). Estramustine Phosphate Sodium. *Drug Intelligence & Clinical Pharmacy*, 18(5): 368-374.

[15] Saha Pijus, Chhanda Debnath, Gervais Bérubé (2013). Steroid-Linked Nitrogen Mustards as Potential Anticancer Therapeutics: A Review. *Journal Steroid Biochemistry*, 137: 271-300.

[16] Romero Alejandro, Ramón Cacabelos, María J. Oset-Gasque, Abdelouahid Samadi, José Marco-Contelles (2013). Novel Tacrine-Related Drugs as Potential Candidates for the Treatment of Alzheimer's Disease. *Bioorganic Medicinal Chemistry Letters*, 23(7): 1916-1922.

[17] Cavalli Andrea, Maria Laura Bolognesi, Anna Minarini, Michela Rosini, Vincenzo Tumiatti, Maurizio Recanatini, Carlo Melchiorre (2008). Multi Target Directed Ligands to Combat Neurodegenerative Diseases. *Journal of Medicinal Chemistry*, 51(3): 347-372.

[18] Nepovimova Eugenie, Elisa Uliassi, Jan Korabecny, Luis Emiliano Peña-Altamira, Sarah Samez, Alessandro Pesaresi, Gregory E. Garcia, Manuela Bartolini, Vincenza Andrisano, Christian Bergamini, Romana Fato, Doriano Lamba, Marinella Roberti, Kamil Kuca, Barbara Monti and Maria Laura Bolognesi (2014). Multitarget Drug Design Strategy: Quinone-Tacrine Hybrids Designed to Block Amyloid-β Aggregation and To Exert Anticholinesterase and Antioxidant Effects. *Journal of Medicinal Chemistry*, 57: 8576-8589.

[19] Howieson Vanessa M., Elisa Tran, Annabelle Hoegl, Han Ling Fam, Jonathan Fu, Kate Sivonen, Xiao Xuan Li, Karine Auclair, Kevin J. Saliba (2019). Triazole Substitution of a Labile Amide Bond Stabilizes Pantothenamides and Improves Their Antiplasmodial Potency. *Antimicrobial Agents Chemotherapy*, 60: 7146-7152.

[20] Divo Alan A., Timothy G. Geary, Nella L. Davis, James B. Jensen (1985). Nutritional requirements of Plasmodium falciparum in culture. I. Exogenously supplied dialyzable components necessary for continuous growth. *Journal of Protozoology*, 32: 59-64.

[21] Spry Christina, Kevin J. Saliba (2009). The human malaria parasite Plasmodium falciparum is not dependent on host coenzyme A biosynthesis. *Journal of Biological Chemistry*, 284: 24904-24913.

[22] Spry Christina, Donelly A. van Schalkwyk, Erick Strauss, Kevin J. Saliba (2010). Pantothenate utilization by Plasmodium as a target for antimalarial chemotherapy. *Infectious Disorders - Drug Targets*,10: 200–216.

[23] Saliba Kevin J., Isabelle Ferru, Kiaran Kirk (2005). Provitamin B5 (pantothenol) inhibits growth of the intraerythrocytic malaria parasite. *Antimicrobial and Agents Chemotherapy*, 49: 632-637.

[24] Hoegl Annabelle, Hamed Darabi, Elisa Tran, Emelia Awuah, Eleanor S. C. Kerdo, Eric Habib, Kevin J. Saliba, Karine Auclair (2014). Stereochemical modification of geminal dialkyl substituents on pantothenamides alters antimicrobial activity. *Bioorganic Medicinal Chemistry Letters*, 24: 3274-3277.

[25] Spry Chistina, Christina L. L. Chai, Kiaran Kirk, Kevin J. Saliba (2005). A class of pantothenic acid analogs inhibits Plasmodium falciparum pantothenate kinase and represses the proliferation of malaria parasites. *Antimicrobial and Agents Chemotherapy*, 49: 4649–4657.

[26] Bozorova Khurshed, Jiangyu Zhao, Haji A. Ais (2019). 1,2,3-Triazole-containing hybrids as leads in medicinal chemistry: A recent overview. *Bioorganic & Medicinal Chemistry*, 27: 3511-3531.

[27] Melo Júlio O. F., Claudio L. Donnici, Rodinei Augusti, Vitor F. Ferreira, Maria Cecília B. V. de Souza, Maria Lourdes G. Ferreira, Anna C. Cunha (2006). Heterociclos 1,2,3-Triazólicos: Histórico, Métodos De Preparação, Aplicações e atividades Farmacológicas. *Química Nova*, 29: 569-579. [1,2,3-Triazole heterocycles: History, Preparation Methods, Applications and Pharmacological activities.]

[28] Wim Dehaen and Vasiliy A. Bakulev, (Eds.), *Chemistry of 1,2,3-triazoles in Topics in Heterocyclic Chemistry,* Springer, Switzerland 2015.

[29] Huisgen Rolf (1963). 1.3-Dipolare cycloadditionen rückschau und ausblick. *Angewandte Chemie*, *75*: 604-637.

[30] Forezi Luana da S. M., Mariana F. C. Cardoso, Daniel T. G. Gonzaga, Fernando de C. da Silva, Vitor F. Ferreira (2018).

Alternative Routes to the Click Method for the Synthesis of 1,2,3-Triazoles, an Important Heterocycle. *Medicinal Chemistry*, 18: 1428-1453.

[31] Faraz Khan M., Verma Garima, Akhtar Wasim, Marella Akranth, Alam M. Mumtaz, Akhter Mymoona, Husain Asif, Hasan S. Misbahul, Shaquiquzzaman Mohammad, Haider S. Rashiduddin (2017). Synthetic Trends Followed for the Development of 1,2,3-Triazole Derivatives. *International Journal of Drug Development and Research*, 9: 22-25.

[32] Huisgen Rolf (1963). Kinetik und mechanismus 1.3-dipolarer cycloadditionen. *Angewandte Chemie*, 75: 742-754.

[33] Huisgen Rolf (1963). Kinetics and mechanism of 1,3-dipolar cycloadditions. *Angewandte Chemie International Edition*, 2: 633-645.

[34] Wei Jinjian, Y. Wang, X. L. Wang. (2011) Recent advances of 1, 2, 3-triazole compounds in medicinal chemistry. *Journal of Chinese Pharmaceutical Sciences,* 46: 481-485.

[35] Zhou C. H., Y. Wang (2012). Recent Researches in Triazole Compounds as Medicinal Drugs. *Current Medicinal Chemistry,* 19: 239 – 280.

[36] Wu Peng, Valery V. Fokin (2007). Catalytic azide-alkyne cycloaddition: reactivity and applications. *Aldrichimica acta*, 40: 7-17.

[37] Zarei Amin, Leila Khazdooz, Abdol R. Hajipour, Hamidreza Aghaei, Ghobad Azizi (2012). Macrowave-assisted click chemistry synthesis of 1,2,3-triazoles from aryldiazonium silica sulfates in water. *Synthesis,* 44: 3353-3360.

[38] Vatmurgea Namdev S., Braja G. Hazra, Vandana S. Pore, Fazal Shirazi, Pradnya S. Chavan, Mukund V. Deshpande (2008). Synthesis and antimicrobial activity of beta-lactam-bile acid conjugates linked via triazole. *Bioorganic & Medicinal Chemistry Letters*, 18: 2043-2047.

[39] Whiting Matthew, John Muldoon, Ying-Chuan Lin, Steven M. Silverman, William Lindstrom, Arthur J. Olson, Hartmuth C. Kolb,

M. G. Finn, K. Barry Sharpless, John H. Elder, Valery V. Fokin (2006). Inhibitors of HIV-1 Protease by Using In Situ Click Chemistry. *Angewandte Chemie International Edition,* 45: 1435-1439.

[40] Kolb Hartmuth C., K. B. Sharpless. (2003). The growing impact of click chemistry on drug discovery. *Drug Discovery Today,* 8: 1128-1137.

[41] Sharma Pratibha, Ashok Kumar, Siya Upadhyay, Jitendra Singh, Vinita Sahu (2010). A novel approach to the synthesis of 1,2,3 triazoles and their SAR studies. *Medicinal Chemistry Research,* 19: 589-602.

[42] Chen, S.; Lou, R.; Wu, Y.; Zhou, J.; Hanselmann, R. *Process for the synthesis of triazoles*, U.S. Patent 2010/0234615 A1, September 16, 2010.

[43] Dheer Divya, Virender Singh, Ravi Shankar (2017). Medicinal attributes of 1,2,3-triazoles: Current developments. *Bioorganic Chemistry,* 71: 30-54.

[44] Angell Yu L., Kevin Burgess (2007). Peptidomimetics via copper-catalyzed azide-alkyne cycloadditions. *Chemical Society Reviews,* 36: 1674-1689.

[45] Pedersen Daniel S., Andrew Abell (2011). 1,2,3-Triazoles in peptidomimetic chemistry. *European Journal of Medicinal Chemistry,* 3: 2399-2411.

[46] Agalave Sandip G., Suleman R. Maujan, Vandana S. Pore (2006). Click chemistry: 1,2,3-triazoles as pharmacophores. *Chemistry- An Asian Journal*, 6: 2696-2718.

[47] Schröder David C., Oliver Kracker, Tanja Fröhr, Jerzy Góra, Michał Jewginski, Anke Nieß, Iris Antes, Rafał Latajka, Antoine Marion, Norbert Sewal (2019). 1,4-Disubstituted 1H-1,2,3-Triazole Containing Peptidotriazolamers: A New Class of Peptidomimetics With Interesting Foldamer Properties. *Frontiers in Chemistry,* 7: 155.

[48] Therrien Christian, Roger C. Levesque (2000). Molecular basis of antibiotic resistance and β-lactamase inhibition by mechanism-based

inactivators: Perspectives and future directions. *FEMS Microbiology Reviews,* 24: 251-262.

[49] Khan Fahmi Y., Abdelhaleem Elhiday, Imran F. Khudair, Hind Yousef, Aisha H. Omran, Samer H. Alsamman, Maha Elhamid (2012). Evaluation of the use of piperacillin/tazobactam (Tazocin®) at Hamad General Hospital, Qatar: Are there unjustified prescriptions? *Infection and Drug Resistance,* 5: 17-21.

[50] Delgado-Valverde Mercedes, Eva Torres, Adoración Valiente-Mendez, Benito Almirante, Silvia Gómez-Zorrilla, Núria Borrell, Juan E. Corzo, Mercedes Gurgui, Manuel Almela, Lara García-Álvarez, María C. Fontecoba-Sánchez, Luis Martínez-Martínez, Rafael Cantón, Julia Praena, Manuel Causse, Belén Gutiérrez-Gutiérrez, Jason A. Roberts, Andras Farkas, Álvaro Pascual, Jesús Rodríguez-Banõ (2016). Impact of the MIC of piperacillin /tazobactam on the outcomefor patients with bacteraemia due to Enterobacteriaceae: the Bacteraemia-MIC project. *Journal of Antimicrobial Chemotherapy*, 71: 521-530.

[51] Zhanel George G., Phillip Chung, Heather Adam, Sheryl Zelenitsky, Andrew Denisuik, Frank Schweizer, Philippe R. S. Lagacé-Wiens, Ethan Rubinstein, Alfred S. Gin, Andrew Walkty, Daryl J. Hoban, Joseph P. Lynch 3rd, James A. Karlowsky (2014). Ceftolozane /tazobactam: a novel cephalosporin/β-lactamase inhibitor combination with activity against multidrug-resistant gram-negative bacilli. *Drugs,* 74: 31-51.

[52] Long Timothy E., Justin T. Williams (2014). Cephalosporins currently in early clinical trials for the treatment of bacterial infections. *Expert Opinion on Investigational Drugs*, 23: 1375–1387.

[53] Blackwell C. Caroline, Earl H. Freimer, Gregory C. Tuke (1976). *In vitro* evaluation of the new oral cephalosporin cefatrizine: Comparison with other cephalosporins. *Antimicrobial Agents and Chemotherapy*, 10: 288-292.

[54] Wheless James W., Blanca Vazquez (2010). Rufinamide: A Novel Broad-Spectrum Antiepileptic Drug. *Epilepsy Currents*, 10: 1-6.

[55] Hakimian Shahin, Andrea Cheng-Hakimian, Gail D. Anderson, John W. Miller (2007). Rufinamide: a new anti-epileptic medication. *Expert Opinion on Pharmacotherapy*, 8: 1931-1940.

[56] Brodie Martin J., William E. Rosenfeld, Blanca Vazquez, Rajesh Sachdeo, Carlos Perdomo, Allison Mann, Santiago Arroyo (2009). Rufinamide for the adjunctive treatment of partial seizures in adults and adolescents: a randomized placebo-controlled trial. *Epilepsia*, 50: 1899-909.

[57] Landmark Cecilie J., Philip N. Patsalos (2010). Drug interactions involving the new second- and third-generation antiepileptic drugs. *Journal Expert Review of Neurotherapeutics*, 10: 119-140.

[58] Balzarini Jan, María-Jesús Pérez-Pérez, Ana San-Félix, Dominique Schols, Carlo-Federico Perno, Anne-Mieke Vandamme, María-Jose Camarasa, Erik De Clercq (1992). 2',5'-Bis-O-(tert-butyldimethylsilyl)-3'-spiro-5''-(4''-amino-1'',2''-oxathiole-2'',2'-dioxide)pyrimidine (TSAO) nucleoside analogues: highlyselective inhibitors of human immunodeficiency virus type 1 that are targeted at the viral reverse transcriptase. *Proceedings of the National Academy of Sciences,* 89: 4392–4396.

[59] Das Kaylan, Joseph D. Bauman, Angela S. Rim, Chhaya Dharia, Arthur Clark, Maria-Jose Camarasa, Jan Balzarin, Eddy Arnold (2011). Crystal structure of tert-butyldimethylsilyl-spiroamino oxathioledioxide-thymine (TSAO-T) in complex with HIV-1 reverse transcriptase (RT) redefines the elastic limits of the non-nucleoside inhibitor-binding pocket. *Journal of Medicinal Chemistry*, 54: 2727-2737.

[60] Velázquez Sonsoles, Alvarez, R., Pérez, C., Federico Gago, Eric De Clercq, Jan Balzarini, & María-José Camarasa. (1998). Regiospecific synthesis and anti-human immuno-deficiency virus activity of novel 5-substituted N-alkylcarbamoyl and N,N-dialkyl carbamoyl 1,2,3-triazole-TSAO analogues. *Antiviral Chemistry & Chemotherapy,* 9: 481-489.

[61] Balzarini Jan, María-Jesús Pérez-Pérez, Ana San-Félix, María-José Camarasa, Bathurst IC, Barr PJ & Eric De Clercq. (1992). Kinetics

of inhibition of human immunodeficiency virus type 1 (HIV-1) reverse transcrip-tase by the novel HIV-1 specific nucleoside analogue 2′,5′-bis-O-(t-butyldimethylsilyl)-3′-spiro-5″-(4″-amino-1″,2″-oxathiole-2″,2″-dioxide) thymine (TSAO-T). *Journal of Biological Chemistry,* 267: 11831-11838.

[62] Camarasa María-José, Ana San-Félix, Sonsoles Velázquez, María-Jesús Pérez-Pérez, Federico Gago, Jan Balzarini (2004). TSAO Compounds: The Comprehensive Story of a Unique Family of HIV1 Specific Inhibitors of Reverse Transcriptase. *Current Topics in Medicinal Chemistry*, 4: 671-686.

[63] Thomas Leo, Matthias Eckhardt, Elke Langkopf, Moh Tadayyon, Frank Himmelsbach and Michael Mark (2008). (R)-8-(3-amino-piperidin-1-yl)-7-but-2-ynyl-3-methyl-1-(4-methyl-quinazolin-2-ylmethyl)-3,7-dihydro-purine-2,6-dione (BI 1356), a novel xanthine-based dipeptidyl peptidase 4 inhibitor, has a superior potency and longer duration of action compared with other dipeptidyl peptidase-4 inhibitors. *Journal of Pharmacology and Experimental Therapeutics,* 325: 10-16.

[64] Corrado Chiara, Anna Maria Flugy, Simona Taverna, Stefania Raimondo, Giuliana Guggino, Rashida Karmali, Giacomo De Leo, Riccardo Alessandro (2012). Carboxyamidotriazole-orotate inhibits the growth of imatinib-resistant chronic myeloid leukaemia cells and modulates exosomes-stimulated angiogenesis. *PLoS One,* 7: e42310.

[65] Olivier Mignen, Christine Brink, Antoine Enfissi, Aditi Nadkarni, Trevor J. Shuttleworth, David R. Giovannucci, Thierry Capiod (2005). Carboxyamidotriazole-induced inhibition of mitochondrial calcium import blocks capacitativecalcium entry and cell proliferation in HEK-293 cells. *Journal of Cell Science* 118, 5615-5623.

[66] Das Manjulika (2018). Carboxyamidotriazole orotate in glioblastoma. *The Lancet Oncology*, 19: e292.

[67] Coulton Steven, Irene François (1989). Synthesis of (5R) (E)-6-[(1-methyl-1,2,3-triazol-4-yl)methylene]-7-oxo-1-azabicyclo[3.2.0]hept-

2-ene-2-carboxylic acid, a novel carbapenem derivative, from 6-aminopenicillanic acid. *Tetrahedron Letters,* 30: 3117-3120.

[68] Eby Paul, Maxwell D. Cummings, Oludotun A. Phillips, David P. Czajkowski, Maya P. Singh, Paul Spevak, Ronald G. Micetich, Samarendra N. Maiti. (1995). β-Lactamase inhibitors: synthesis and in vitro evaluation of 6-[(1-substituted-1,2,3-triazol-4-yl) methylene]-penicillanic acid sulfones. *Heterocycles*, 42: 663-668.

[69] Zhang Yansheng, Vincent A. Steingrube, Richard J. Wallace Jr. (1992). Beta-Lactamase Inhibitors and the Inducibility of the Beta-Lactamase of Mycobacterium tuberculosis. *American Journal of Respiratory and Critical Care Medicine*, 145: 657-660.

[70] Boechat Núbia, Maria de L. G. Ferreira, Luiz C. S. Pinheiro, Antônio M. L. Jesus, Milene M. M. Leite, Anna C. C. Aguiar, Isabel M de Andrade, Antoniana U. Krettli (2014). New Compounds Hybrids 1H-1,2,3-Triazole-Quinoline Against Plasmodium Falciparum. *Chemical Biology & Drug Design,* 84: 325-32.

[71] Santos Juliana de O., Guilherme R. Pereira, Geraldo C. Brandão, Tatiane F. Borgati, Lucas M. Arantes, Renata C. de Paula, Luciana F. Soares, Maria F. A. do Nascimento, Márlia R. C. Ferreira, Alex G. Taranto, Fernando P. Varottig, Alaíde B. de Oliveira (2016). Synthesis, in vitro Antimalarial Activity and in silico Studies of Hybrid Kauranoid 1,2,3-Triazoles Derived from Naturally Occurring Diterpenes. *Journal of the Brazilian Chemical Society,* 27: 551-565.

[72] Brandão Geraldo C., Franciele C. R. Missias, Lucas M. Arantes, Luciana F. Soares, Kuldeep K. Roy, Robert J. Doerksen, Alaide B. de Oliveira, Guilherme R. Pereira (2018). Antimalarial naphthoquinones Synthesis via click chemistry, in vitro activity, docking to PfDHODH and SAR of lapachol-based compounds. *European Journal of Medicinal Chemistry*, 145: 191-205.

[73] Chua Xue-Mei, Cong Wang, Wen-Ling Wang, Li-Li Liang, Wen Liu, Kai-Kai Gong, Kun-LaiSun (2019). Triazole derivatives and their antiplasmodial and antimalarial activities. *European Journal of Medicinal Chemistry,* 166: 206-223.

[74] Manohar Sunny, Shabana I. Khan, Diwan S. Rawat (2011). Synthesis of 4-aminoquinoline-1,2,3-triazole and 4-aminoquinoline-1,2,3-triazole-1,3,5-triazine Hybrids as Potential Antimalarial Agents. *Chemical Biology & Drug Design,* 78: 124-136.

[75] Batra Neha, Vinoth Rajendran, Drishti Agarwal, Ishan Wadi, Prahlad C. Ghosh, Rinkoo D. Gupta, Mahendra Nath (2018). Synthesis and Antimalarial Evaluation of [1, 2,3]-Triazole-Tethered Sulfonamide-Berberine Hybrids. *ChemistrySelect*, 3: 9790-9793.

[76] Kapkoti Deepak S., Shilpi Singh, Suaib Luqmanb, Rajendra S. Bhakuni (2018). Synthesis of novel 1,2,3-triazole based artemisinin derivatives and their antiproliferative activity. *New Journal of Chemistry*, 42: 5978-5995.

[77] D'hooghe Matthias, Stéphanie Vandekerckhove, Karen Mollet, Karel Vervisch, Stijn Dekeukeleire, Liesbeth Lehoucq, Carmen Lategan, Peter J. Smith, Kelly Chibale, Norbert De Kimpe (2011). Synthesis of 2-amino-3-arylpropan-1-ols and1-(2,3-diaminopropyl)-1,2,3-triazoles and evaluation of their antimalarial activity. *Beilstein Journal of Organic Chemistry*, 7: 1745–1752.

[78] Vanessa M. Howieson, Elisa Tran, Annabelle Hoegl, Han L. Fam, Jonathan Fu, Kate Sivonen, Xiao X. Li, Karine Auclair, Kevin J. Saliba (2016). Triazole substitution of a labile amide bond stabilizes pantothenamides and improves their antiplasmodial potency. *Antimicrobial Agents and Chemotherapy,* 60: 7146-7152.

[79] Lebeau Alexandre, Cyril Abrioux, David Bénimèlis, Zohra Benfodda, Patrick Meffre (2017). Synthesis of 1,4-disubstituted 1,2,3-triazole Derivatives Using Click Chemistry and their Src Kinase Activities. *Medicinal Chemistry Medicinal Chemistry*, 13: 40-48.

[80] Borgati Tatiane F., Guilherme R. Pereira, Geraldo C. Brandão, Juliana O. Santos, Dayane Aparecida M. Fernandes, Renata C. de Paula, Maria Fernanda A. do Nascimento, Luciana F. Soares, Júlio César D. Lopes, José D. de Souza Filho, Alaíde B. de Oliveira (2017). Synthesis by Click Reactions and Antiplasmodial Activity of

Lupeol 1,2,3-Triazole Derivatives. *Journal of the Brazilian Chemical Society,* 28: 1850-1856.

[81] Balabadra Saikrishn, MeenaKumari Kotni, Vijjulatha Manga, Aparna D. Allanki, Rajesh Prasad, Puran S. Sijwali. (2017). Synthesis and evaluation of naphthyl bearing 1,2,3-triazole analogs as antiplasmodial agents, cytotoxicity and docking studies. *Bioorganic & Medicinal Chemistry*, 25: 221-232.

[82] Soares Roberta R., Roberta C. N. R. Corrales, Fernanda V. Lopes, Marcio J. M. Alves, Adilson D. da Silva, Clarice Abramo (2013). Synthesis and *In Vivo* Antimalarial Activity of Novel Derivatives of 6- Mercaptopurine. *Letters in Drug Design & Discovery,* 10: 741-747.

[83] Taleli Lebusetsa, Carmende Kock, Peter J. Smith, Stephen C. Pelly, Margaret A. L. Blackie, Willem A. L.van Otterloa (2015). In vitro antiplasmodial activity of triazole-linked chloroquinoline derivatives synthesized from 7-chloro-N-(prop-2-yn-1-yl)quinolin-4-amine. *Bioorganic & Medicinal Chemistry,* 23: 4163-4171.

[84] Batra Neha, Vinoth Rajendran, Drishti Agarwal, Ishan Wadi, Prahlad C. Ghosh, Rinkoo D. Gupta and Mahendra Nath (2019). Synthesis and Antimalarial Evaluation of [1,2,3]-Triazole-Tethered Sulfonamide-Berberine Hybrids. *Chemistry Select*, 3: 9790 –9793.

[85] Aufort Marie, Jean Herscovici, Pascale Bouhours, Nicole Moreau, Christian Girard (2008). Synthesis and antibiotic activity of a small molecules library of 1,2,3-triazole derivatives. *Bioorganic & Medicinal Chemistry Letters,* 18: 1195-1198.

[86] Pereira David, Prabhavathi Fernandes (2011). Synthesis and antibacterial activity of novel 4-aryl-[1,2,3]-triazole containing macrolides. *Bioorganic & Medicinal Chemistry Letters*, 21: 510-513.

[87] Phillips Oludotun A., Edet E. Udo, Mohammed E. Abdel-Hamid, RenyVarghese (2009). Synthesis and antibacterial activity of novel 5-(4-methyl-1H-1,2,3-triazole) methyl oxazolidinones. *European Journal of Medicinal Chemistry*, 44: 3217-3227.

[88] López-Rojas Priscila, Monika Janeczko, Konrad Kubiński, Ángel Amesty, Maciej Masłyk, Ana Estévez-Braun (2018). Synthesis and Antimicrobial Activity of 4-Substituted 1,2,3-Triazole-Coumarin Derivatives. *Molecules*, 23: 199.

[89] Al-blewi Fawzia Faleh, Meshal A. Almehmadi, Mohamed R. Aouad, Sanaa K. Bardaweel, Pramod K. Sahu, Mouslim Messali, Nadjet Rezki, El Sayed H. El Ashry (2018). Design, synthesis, ADME prediction and pharmacological evaluation of novel benzimidazole-1,2,3-triazole-sulfonamide hybrids as antimicrobial and antiproliferative agents. *Chemistry Central Journal 12*, Article number: 110.

[90] Wang Xian-Long, Kun Wan, Cheng-He Zhou (2010). Synthesis of novel sulfanilamide-derived 1,2,3-triazoles and their evaluation for antibacterial and antifungal activities. *European Journal of Medicinal Chemistry*, 45: 4631-4639.

[91] Lima-Neto Reginaldo G., Nery N. M. Cavalcante, Rajendra M. Srivastava, Francisco J. B. Mendonça Junior, Almir G. Wanderley, Rejane P. Neves, Janaína V. dos Anjos (2010). Synthesis of 1,2,3-Triazole Derivatives and *in Vitro* Antifungal Evaluation on *Candida* Strains. *Molecules*, 17: 5882-5892.

[92] Kategaonkar Amol H., Pravin V. Shinde, Atul H. Kategaonkar, Sharad K. Pasale, Bapurao B. Shingate, Murlidhar S. Shingare (2010). Synthesis and biological evaluation of new 2-chloro-3-((4-phenyl-1H-1,2,3-triazol-1-yl)methyl)quinoline derivatives via click chemistry approach. *European Journal of Medicinal Chemistry*, 45: 3142-3146.

[93] Shalini Kumari, Nitin Kumar, Sushma Drabu, Pramod K. Sharma1 (2011). Advances in synthetic approach to and antifungal activity of traizoles. *Beilstein Journal of Organic Chemistry*, 7: 668-677.

[94] Daia Zhi-Cheng, Yong-Fei Chena, Mao Zhange, Sheng-Kun Lia, Ting-Ting Yanga, Li Shen, Jian-Xin Wang, Shao-Song Qian, Hai-Liang Zhu, Yong-Hao Ye (2015). Synthesis and antifungal activity of 1,2,3-triazole phenylhydrazone derivatives. *Organic Biomolecular Chemistry*, 13: 477-486.

[95] da Silva Iara F., Prisicila R. C. Martins, Emanuelly G. da Silva, Sabrina B. Ferreira, Vitor F. Ferreira, Karen R. C. da Costa, Marne C. de Vasconcellos, Emerson S. Lima, Fernando de C. da Silva (2013). Synthesis of 1H-1,2,3-triazoles and Study of their Antifungal and Cytotoxicity Activities. *Medicinal Chemistry,* 9: 1085-1090.

[96] Boechat Nubia, Vitor F. Ferreira, Sabrina B. Ferreira, Maria de Lourdes G. Ferreira, Fernando de C. da Silva, Monica M. Bastos, Marilia dos S. Costa, Maria Cristina S. Lourenço, Angelo C. Pinto, Antoniana U. Krettli, Anna Caroline Aguiar, Brunno M. Teixeira, Nathalia V. da Silva, Priscila R. C. Martins, Flavio Augusto F. M. Bezerra, Ane Louise S. Camilo, Gerson P. da Silva, Carolina C. P. Costa (2011). Novel 1,2,3-triazole derivatives for use against Mycobacterium tuberculosis H37Rv (ATCC 27294) strains. *Journal of Medicinal Chemistry*, 54 : 5988-99.

[97] Gallardo Hugo, Gilmar Conte, Fernando Bryk, Maria C. S. Lourenço, Marilia S. Costa, Vitor F. Ferreira (2007). Synthesis and evaluation of 1-Alkyl-4- phenyl-[1,2,3]triazole derivatives as antimycobacterial agent. *Journal of the Brazilian Chemical Society*, 18: 1285-91.

[98] Surineni Goverdhan, Perumal Yogeeswari, Dharmarajan Sriram, Srinivas Kantevari (2015). Rational design, synthesis and evaluation of novel-substituted 1,2,3-triazolylmethyl carbazoles as potent inhibitors of Mycobacterium tuberculosis, *Medicinal Chemistry Research*, 24: 1298-1309.

[99] Shaikh Mubarak H., Dnyaneshwar D. Subhedar, Laxman Nawale, Dhiman Sarkar, Firoz A. Kalam Khan, Jaiprakash N. Sangshetti, Bapurao B. Shingate (2015). 1,2,3-Triazole derivatives as antitubercular agents: synthesis, biological evaluation and molecular docking study. *Medicinal Chemistry Communications*, 6: 1104-1116.

[100] Keri Rangappa S., Siddappa A. Patil, Srinivasa Budagumpi, Bhari M. Nagaraja (2015). Triazole: A Promising Antitubercular Agent. *Chemical Biology Drug Design,* 86: 410-423.

[101] Nalla Viswanadh, Aslam Shaikh, Sanket Bapat, Renu Vyas, M. Karthikeyan, P. Yogeeswari, D. Sriram, M. Muthukrishnan (2018). Identification of potent chromone embedded [1,2,3]-triazoles as novel anti-tubercular agents. *Royal Society Open Science*, 5: 171750.

[102] Phatak Pramod S., Rajubai D. Bakale, Sambhaji T. Dhumal, Lalita K. Dahiwade, Prafulla B. Choudhari, Vagolu Siva Krishna, Dharmarajan Sriram, Kishan P. Haval (2019). Synthesis, antitubercular evaluation and molecular docking studies of phthalimide bearing 1,2,3-triazoles. *Synthetic Communications,* 49: 2017-2028.

[103] Jordão Alessandro K., Plínio C. Sathler, Vitor F. Ferreira, Vinícius R. Campos, Maria C. B. V. de Souza, Helena C. Castro, Andressa Lannes, André Lourenco, Carlos R. Rodrigues, Murilo L. Bello, Maria C. S. Lourenco, Guilherme S. L. Carvalho, Maria C. B. Almeida, Anna C. Cunha (2011). Synthesis, antitubercular activity, and SAR study of *N*-substituted-phenylamino-5-methyl-1*H*-1,2,3-triazole-4-carbohydrazides. *Bioorganic & Medicinal Chemistry*, 19: 5605-5611.

[104] Ferreira Marcelle de L., Marcus V. N. de Souza, Solange M. S. V. Wardell, James L. Wardell, Thatyana R. A. Vasconcelos, Vitor F. Ferreira, Maria C. S. Lourenço (2010). Synthesis and antitubercular evaluation of new bis-1,2,3-triazoles derived from D-mannitol. *Journal of Carbohydrate Chemistry,* 29: 265-274.

[105] Conte Mauro Lo, Alberto Marra, Angela Chambery, Sudagar S. Gurcha, Gurdyal S. Besra, Alessandro Dondoni (2010). Modular Approach to Triazole-Linked 1,6-α-D-Oligomannosides to the Discovery of Inhibitors of Mycobacterium tuberculosis Cell Wall Synthetase. *Journal of Organic Chemistry*, 75: 6326-6336.

[106] Kumar Kewal, Séverine Carrère-Kremer, Laurent Kremer, Yann Guérardel, Christophe Biot, Vipan Kumar (2013). 1H-1,2,3-Triazole-Tethered Isatin−Ferrocene and Isatin−Ferrocenylchalcone Conjugates: Synthesis and in VitroAntitubercular Evaluation. *Organometallics*, 32: 5713−5719.

[107] Akolkar Satish V., Amol A. Nagargoje, Vagolu S. Krishna, Dharmarajan Sriram, Jaiprakash N. Sangshetti, Manoj Damaled, Bapurao B. Shingate (2019). New N-phenylacetamide-incorporated 1,2,3-triazoles: [Et_3NH][OAc]-mediated efficient synthesis and biological evaluation. *RSC Advances*, 9: 22080-22091.

[108] Zhou Longhu, Adel Amer, Michael Korn, Robert Burda, Jan Balzarini, Erik De Clercq, Earl R. Kern, Paul F. Torrence (2005). Synthesis and antiviral activities of 1,2,3-triazole functionalized thymidines: 1,3-dipolar cycloaddition for efficient regioselective diversity generation. *Antiviral Chemistry & Chemotherapy*, 16: 375-383.

[109] da Silva Fernando de C., Maria C. B. V. de Souza, Izabel I. P. Frugulhetti, Helena C. Castro, Silmara L. de O. Souza, Thiago M. L. de Souza, Diego Q. Rodrigues, Alessandra M. T. Souza, Paula A. Abreu, Fabiana Passamani, Carlos R. Rodrigues, Vitor F. Ferreira (2009). Synthesis, HIV-RT inhibitory activity and SAR of 1-benzyl-1*H*-1,2,3-triazole derivatives of carbohydrates. *European Journal of Medicinal Chemistry,* 44: 373-383.

[110] Whiting Matthew, Jonathan C. Tripp, Ying-Chuan Lin, William Lindstrom, Arthur J. Olson, John H. Elder K. Barry Sharpless, Valery V. Fokin (2006). Rapid Discovery and Structure-Activity Profiling of Novel Inhibitors of Human Immunodeficiency Virus Type 1 Protease Enabled by the Copper(I)-Catalyzed Synthesis of 1,2,3-Triazoles and Their Further Functionalization. *Journal of Medicinal Chemistry*, 49: 7697-7710.

[111] Gonzaga Daniel T. G., Thiago M. L. Souza, Viviane M. M. Andrade, Vitor F. Ferreira, Fernando de C. da Silva (2018). Identification of 1-Aryl-1H-1,2,3-triazoles as Potential New Antiretroviral Agents. *Medicinal Chemistry*, 14: 242-248.

[112] El-Sayed W. A., H. S. Khalaf, S. F. Mohamed, H. A. Hussien, O. M. Kutkatc, A. E. Amrd (2017). Synthesis and Antiviral Activity of 1,2,3-Triazole Glycosides Based Substituted Pyridine via Click Cycloaddition. *Russian Journal of General Chemistry,* 87: 2444-2453.

[113] Ferreira Maria de Lourdes G., Luiz C. S. Pinheiro, Osvaldo A. Santos-Filho, Marta D. S. Peçanha, Carolina Q. Sacramento, Viviane Machado, Vitor F. Ferreira, Thiago Moreno L. Souza, Núbia Boechat (2014). Design, synthesis, and antiviral activity of new 1H-1,2,3-triazolenucleoside ribavirin analogs. *Medicinal Chemistry Research,* 23: 1501-1511.

[114] Youssif Bahaa G. M., Yaseen A. M. Mohamed, Mohammed T. A. Salim, Fuyuhiko Inagaki, Chisato Mukai, Hajjaj H. M. Abdu-Allah (2016). Synthesis of some benzimidazole derivatives endowed with 1,2,3-triazole as potential inhibitors of hepatitis C virus. *Acta Pharmaceutica*, 66: 219-231.

[115] Jordão Alessandro K., Vitor F. Ferreira, Thiago M. L. Souza, Gabrielle G. de Souza Faria, Viviane Machado, Juliana L. Abrantes, Maria C. B. V. de Souza, Anna C. Cunha (2011). Synthesis and anti-HSV-1 activity of new 1,2,3-triazole derivatives. *Bioorganic & Medicinal Chemistry*, 19: 1860-1865.

[116] Alessandro K. Jordão, Priscila P. Afonso, Vitor F. Ferreira, Maria C. B. V. de Souza, Maria C. B. Almeida, Cristiana O. Beltrame, Daniel P. Paiva, Solange M. S. V. Wardell, James L. Wardell, Edward R. T. Tiekink, Clarissa R. Damaso, Anna C. Cunha (2009). Antiviral evaluation of N-amino-1,2,3-triazoles against Cantagalo virus replication in cell culture. *European Journal of Medicinal Chemistry*, 44: 3777-3783.

[117] Ferreira Sabrina B., Marilia S. Costa, Núbia Boechat, Rômulo J. S. Bezerra, Marcelo S. Genestra, Marilene M. Canto-Cavalheiro, Warner B. Kover, Vitor F. Ferreira (2007). Synthesis and evaluation of new difluoromethyl azoles as antileishmanial agents. *European Journal of Medicinal Chemistry*, 42: 1388-1395.

[118] Cassamale Tatiana B., Eduarda C. Costa, Diego B. Carvalho, Nadla S. Cassemiro, Carolina C. Tomazela, Maria C. S. Marques, Mariáh Ojeda, Maria F. C. Matos, Sérgio Albuquerque, Carla C. P. Arruda, Adriano C. M. Baroni (2016). Synthesis and Antitrypanosomastid Activity of 1,4-Diaryl-1,2,3-triazole Analogues of Neolignans

Veraguensin, Grandisin and Machilin G. *Journal of the Brazilian Chemical Society*, 27: 1217-1228.

[119] Maji Krishnendu, Mazharul Abbasi, Debasish Podder, Rupak Datta, Debasish Haldar (2018). Potential Antileishmanial Activity of a Triazole-Based Hybrid Peptide against *Leishmania major. ChemistrySelect,* 3: 10220-10225.

[120] Meinel Raíssa S., Ayla das C. Almeida, Pedro H. F. Stroppa, Nícolas Glanzmann, Elaine S. Coimbra, Adilson D. da Silva (2019). Novel functionalized 1,2,3-triazole derivatives promote antileishmanial activity, increase in total and mitochondrial-ROS and depolarization of mitochondrial membrane potential of Leishmania amazonensis. *Chemico-Biological Interactions* 108850, in Press.

[121] Porta Exequiel O. J., Sebastián N. Jäger, Isabel Nocito, Galina I. Lepesheva, Esteban C. Serra, Babu L. Tekwanid, Guillermo R. Labadie (2017). Antitrypanosomal and antileishmanial activity of prenyl-1,2,3-triazoles. *Medicinal Chemistry Communicatons,* 8: 1015-1021.

[122] Oliveira Vanessa G., Viviane dos S. Faiões, Gabrieli B. R. Gonçalves, Miriam F. O. Lima, Fernanda C. S. Boechat, Anna C. Cunha, Valter V. de Andrade-Neto, Fernando de C. da Silva, Eduardo C. Torres-Santos, Maria Cecília B. V. de Souza (2018). Design, Synthesis and Antileishmanial Activity of Naphthotriazolyl-4- Oxoquinolines. *Current Topics in Medicinal Chemistry*, 18: 1454-1464.

[123] Raj Raghu, Pardeep Singh, Parvesh Singh, Jiri Gut, Philip J. Rosenthal, Vipan Kumar (2013). Azide-alkyne cycloaddition en route to 1H-1,2,3-triazole-tethered 7-chloroquinoline-isatin chimeras: Synthesis and antimalarial Evaluation. *European Journal of Medicinal Chemistry,* 62: 590-596.

[124] Teixeira Robson R., Poliana A. R. Gazolla, Adalberto M. da Silva, Maria P. G. Borsodi, Bartira R. Bergmann, Rafaela S. Ferreira, Boniek G. Vaz, Gessica A. Vasconcelos and Wallace P. Lima (2018). Synthesis and leishmanicidal activity of eugenol derivatives

bearing 1,2,3-triazole functionalities. *European Journal of Medicinal Chemistry*, 146: 274-286.

[125] Leaver David J., Raymond M. Dawson, Jonathan M. White, Anastasios Polyzos, Andrew B. Hughes (2011). Synthesis of 1,2,3-triazole linked galactopyranosides and evaluation of cholera toxin inhibition. *Organic & Biomolecular Chemistry,* 9: 8465.

[126] Pokhodylo Nazariy, Olga Shyyka, Vasyl Matiychuk (2013). Synthesis of 1,2,3-Triazole Derivatives and Evaluation of their Anticancer Activity. *Scientia Pharmaceutica*, 81: 663-676.

[127] Zhang Wenjuan, Zhi Li, Meng Zhou, Feng Wu, Xueyan Hou, Hao Luo, Hao Liu, Xuan Han, Guoyi Yan, Zhenyu Ding, Rui Li (2014). Synthesis and biological evaluation of 4-(1,2,3-triazol-1-yl)coumarin derivatives as potential antitumor agents. *Bioorganic & Medicinal Chemistry Letters,* 24: 799-807.

[128] Rao P. Sambasiva, C. Kurumurthy, B. Veeraswamy, G. Santhosh Kumara, Y. Poornachandra, C. Ganesh Kumar, Sathish Babu Vasamsetti, Srigiridhar Kotamraju, B. Narsaiaha (2014). Synthesis of novel 1,2,3-triazole substituted-N-alkyl/aryl nitrone derivatives, their anti-inflammatory and anticancer activity. *European Journal of Medicinal Chemistry*, 80: 184-191.

[129] Prachayasittikul Veda, Ratchanok Pingaew, Nuttapat Anuwongcharoen, Apilak Worachartcheewan, Chanin Nantasenamat, Supaluk Prachayasittikul, Somsak Ruchirawat, Virapong Prachayasittikul (2015). *Discovery of novel 1,2,3-triazole derivatives as anticancer agents using QSAR and in silico structural modification.* Springer Plus 4, Article number: 571.

[130] Branco Jessica R., Vanessa G. Oliveira, Amanda M. Esteves, Ingrid C. Chipoline, Miriam F. O. Lima, Fernanda C.S. Boechat, Fernando C. da Silva, Vítor F. Ferreira, Mauro Sola-Penna, Maria C. B. V. de Souza, Patricia Zancan (2018). A Novel Naphthotriazolyl-4-oxoquinoline Derivative that Selectively Controls Breast Cancer Cells Survival Through the Induction of Apoptosis. *Current Topics in Medicinal Chemistry*, 18: 1465 – 1474.

[131] Reis Wallace J., Paulo O. L. Moreira, Rosemeire B. Alves, Heloísa H. M. Oliveira, Luciana M. Silva, Fernando P. Varotti and Rossimiriam P. Freitas (2018). Novel Symmetrical 1,4-Disubstituted-bis-1,2,3-Triazoles: Synthesis by Double CuAAC and Cytotoxicity Evaluation. *Current Topics in Medicinal Chemistry* 18: 1475-1482.

[132] Aouad Mohamed R., Moataz A. Soliman, Muath O. Alharbi, Sanaa K. Bardaweel, Pramod K. Sahu, Adeeb A. Ali, Mouslim Messali, Nadjet Rezki, Yaseen A. Al-Soud (2018). Design, Synthesis and Anticancer Screening of Novel Benzothiazole-Piperazine-1,2,3-Triazole Hybrids. *Molecules,* 23: 2788,

[133] Filho Ricardo I., Daniel T. G. Gonzaga, Thainá M. Demaria, João G. B. Leandro, Dora C. S. Costa, Vitor F. Ferreira, Mauro Sola-Penna, Fernando de C. da Silva, Patricia Zancan (2018). A Novel Triazole Derivative Drug Presenting In Vitro and In Vivo Anticancer Properties. *Current Topics in Medicinal Chemistry*, 18: 1483 – 1493.

[134] Dong Hong-Ru, Jian-Guo Wu. Design (2018). Synthesis and anticancer activity evaluation of aziridine-1,2,3-triazole hybrid derivatives. *Heterocycle Communication*, 24: 109-112.

[135] de O. Assis Shalom P., Moara T. da Silva, Ronaldo N. de Oliveira, Vera Lucia de M. Lima (2012). Synthesis and Anti-Inflammatory Activity of New Alkyl-Substituted Phthalimide 1H-1,2,3-Triazole Derivatives. *The Scientific World Journal*, ID 925925.

[136] Wuest Frank, Xinli Tang, Torsten Kniess, Jens Pietzsch, Mavanur Suresh (2009). Synthesis and cyclooxygenase inhibition of various (aryl-1,2,3-triazole-1-yl)-methanesulfonylphenyl derivatives. *Bio-organic & Medicinal Chemistry,* 17: 1146-51.

[137] Shafi Syed, Mohammad Mahboob Alam, Naveen Mulakayala, Chaitanya Mulakayala, G. Vanajad, Arunasree M. Kalle, Reddanna Pallu, M. S. Alama (2012). Synthesis of novel 2-mercapto benzothiazole and 1,2,3-triazole based bis-heterocycles: Their anti-inflammatory and anti-nociceptive activities. *European Journal of Medicinal Chemistry,* 49: 324-333.

[138] Naaz Fatima, M. C. Preeti Pallavi, Syed Shafi, Naveen Mulakayala, M. Shahar Yar, H. M. Sampath Kumarb (2018). 1,2,3-triazole tethered Indole-3-glyoxamide derivatives as multiple inhibitors of 5-LOX, COX-2 & tubulin: Their anti-proliferative & anti-inflammatory activity. *Bioorganic Chemistry*, 81: 1-20.

[139] Yoon Jieun, Lan Cho, Sang Kook Lee, Jae-Sang Ryu (2011). Syntheses of 1,2,3-triazolyl salicylamides with inhibitory activity on lipopolysaccharide-induced nitric oxide production. *Bioorganic & Medicinal Chemistry Letters*, 21: 1953-1957.

[140] de O. Assis Shalom P., Moara T. da Silva, Filipe T. da Silva, Mirella P. Sant'Anna, Carolina M. B. de A. Tenório, Caroline F. B. dos Santos, Caíque S. M. da Fonseca, Gustavo Seabra, Vera L. M. Lima, Ronaldo N. de Oliveira (2019). Design and Synthesis of Triazole-Phthalimide Hybrids with Antiinflammatory Activity. *Chemical and Pharmaceutical Bulletin,* 67: 96-105.

[141] Angajala Kishore K., Sunitha Vianala, Ramesh Macha, M. Raghavender, Murali K. Thupurani, P. J. Pathi (2016). *Synthesis, anti-inflammatory, bactericidal activities and docking studies of novel 1,2,3-triazoles derived from ibuprofen using click chemistry.* SpringerPlus 5, Article number: 423.

[142] Dasari Srinivasa R., Subbaiah Tondepu, Lakshmana R. Vadali, Nareshvarma Seelam (2019). Design, Synthesis and Molecular Modeling of Nonsteroidal Anti-inflammatory Drugs Tagged Substituted 1,2,3-Triazole Derivatives and Evaluation of Their Biological Activities. *Journal Heterocyclic Chemistry,* 56: 1318-1329.

[143] Angajala, K. K., Vianala, S., Macha, R., Raghavender, M., Thupurani, M. K., & Pathi, P. J. (2016). *Synthesis, anti-inflammatory, bactericidal activities and docking studies of novel 1,2,3-triazoles derived from ibuprofen using click chemistry.* SpringerPlus 5, Article number: 423.

[144] Cunha Anna C., Juliana M. Figueiredo, Jorge L. M. Tributino, Ana L. P. Miranda, Helena C. Castro, Russolina B. Zingali, Carlos A. M. Fraga, Maria Cecília B. V. de Souza, Vitor F. Ferreira, Eliezer J.

Barreiro (2003). Antiplatelet properties of novel *N*-substituted-phenyl-1,2,3-triazole-4-acylhydrazone derivatives. *Bioorganic & Medicinal Chemistry*, 11: 2051-2059.

[145] Jordão Alessandro K., Vitor F. Ferreira, Emerson S. Lima, Maria C. B. V. de Souza, Eduardo C. L. Carlos, Helena C. Castro, Reinaldo B. Geraldo, Carlos R. Rodrigues, Maria C. B. Almeida, Anna C. Cunha (2009). Synthesis, antiplatelet and in silico evaluations of novel *N*-substituted-phenylamino-5-methyl-1*H*-1,2,3-triazole-4-carbohydrazides. *Bioorganic & Medicinal Chemistry*, 17: 3713-3719.

[146] Moura Laura de A., Ana C. M. de Almeida, Andreza V. da Silva, Vivian R. de Souza, Vitor F. Ferreira, Michel V. Menezes, Carlos R. Kaiser, Sabrina B. Ferreira, André L. Fuly (2016). Synthesis, Anticlotting and Antiplatelet Effects of 1,2,3-Triazoles Derivatives. *Medicinal Chemistry*, 12: 733-741.

[147] da Silva Eufrânio N. Júnior, Maria Aline B. F. de Moura, Antonio V. Pinto, Maria do Carmo F. R. Pinto, Maria Cecília B. V. de Souza, Ana J. Araújo, Claudia Pessoa, Letícia V. Costa-Lotufo, Raquel C. Montenegro, Manoel Odorico de Moraes, Vitor F. Ferreira, Marilia O. F. Goulart (2009). Cytotoxic, trypanocidal activities and physicochemical parameters of nor-β-lapachone-based 1,2,3-triazoles. *Journal of the Brazilian Chemical Society*, 20: 635-643.

[148] Bakunov Stanislav A., Svetlana M. Bakunova, Tanja Wenzler, Maedot Ghebru, Karl A. Werbovetz, Reto Brun Richard R. Tidwell (2010). Synthesis and Antiprotozoal Activity of Cationic 1,4-Diphenyl-1H-1,2,3-triazoles. *Journal of Medicinal Chemistry*, 53: 254-272.

[149] Faria Robson X., Daniel T. G. Gonzaga, Paulo A. F. Pacheco, André L. A. Souza, Vitor F. Ferreira, Fernando de C. da Silva (2018). Searching for new drugs for Chagas diseases: triazole analogs display high in vitro activity against Trypanosoma cruzi and low toxicity toward mammalian cells. *Journal of Bioenergetics and Biomembranes*, 50: 81-91.

[150] Giroud Maude, Bernd Kuhn, Sarah Saint-Auret, Christoph Kuratli, Rainer E. Martin, Franz Schuler, François Diederich, Marcel Kaiser, Reto BrunTanja Schirmeister, Wolfgang Haap (2018). 2H-1,2,3-Triazole-Based Dipeptidyl Nitriles: Potent, Selective, and Trypanocidal Rhodesain Inhibitors by Structure-Based Design. *Journal of Medicinal Chemistry,* 61: 3370-3388.

[151] Silva Bianca N. M., Policarpo A. Sales Junior, Alvaro J. Romanha, Silvane M. F. Murta, Camilo H.S. Lima, Magaly G. Albuquerque, Eliane D'Elia, José G. A. Rodrigues, Vitor F. Ferreira, Fernando C. Silva, Angelo C. Pinto, Bárbara V. Silva (2019). Synthesis of New Thiosemicarbazones and Semicarbazones Containing the 1,2,3-1H-triazole-isatin Scaffold: Trypanocidal, Cytotoxicity, Electrochemical Assays, and Molecular Docking. *Medicinal Chemistry*, 15: 240-256.

[152] de Souza Thiago B., Ivo S. Caldas, Favero R. Paula, Camila C. Rodrigues, Diogo T. Carvalho, Danielle F. Dias (2019). Synthesis, activity, and molecular modeling studies of 1,2,3-triazole derivatives from natural phenylpropanoids as new trypanocidal agents. *Chemical Biology & Drug Design,* 00: 1-6.

[153] Leite Débora I., Fábio de V. Fontes, Monica M. Bastos, Lucas V. B. Hoelz, Maria da C. A. D. Bianco, Andressa P. de Oliveira, Patricia B. da Silva, Cristiane F. da Silva, Denise da G. J. Batista, Aline N. S. da Gama, Raiza B. Peres, Jose D. F. Villar, Maria de N. C. Soeiro and Nubia Boechat (2018). New 1,2,3-triazole-based analogues of benznidazole for use against Trypanosoma cruzi infection: In vitro and in vivo evaluations. *Chemical Biology & Drug Design*, 92: 1670-1682.

[154] Biagi, G.; Dell'Omodarme, G.; Ferretti, M.; Giorgi, I.; Livi, O.; Scartoni, V. Structure-activity studies on a 1,2,3-triazole derivative, a potent in vitro inhibitor of prostaglandin synthesis: the role of the heterocyclic ring. *Farmaco* 1992, 47: 335-344.

[155] Biagi, G.; Ferretti, M.; Giorgi, I.; Livi, O.; Scartoni, V.; Lucacchini, A. Studies on 1,2,3-triazole derivatives as in vitro inhibitors of prostaglandin synthesis. *Farmaco* 1993, 48, 1159-1165.

[156] Buckle Derek R., Caroline J. M. Rockell, Harry Smith, Barbara A. Spicer (1984). Studies on 1,2,3,-triazoles. 10. Synthesis and antiallergic properties of 9-oxo-1H,9H-benzothiopyrano[2,3-d]-1,2,3-triazoles and their S-oxides. *Journal of Medicinal Chemistry,* 27: 223-227.

[157] Buckle Derek R., Caroline J. M. Rockell, Harry Smith, Barbara A. Spicer (1986). Studies on 1,2,3-triazoles. 13 (Piperazinylalkoxy) [1]benzopyrano[2,3-d]-1,2,3-triazol-9(1H)-ones with combined H1-antihistamine and mast cell stabilizing properties. *Journal of Medicinal Chemistry,* 29: 2262-2267.

[158] Celik Fatih, Yasemin Unver, Burak Barut, Arzu Ozel, Kemal Sancak (2018). Synthesis, Characterization and Biological Activities of New Symmetric Bis-1,2,3-Triazoles with Click Chemistry. *Medicinal Chemistry*, 14: 230-241.

[159] Bębenek Ewa, Maria Jastrzębska, Monika Kadela-Tomanek, Elwira Chrobak, Beata Orzechowska, Katarzyna Zwolińska, Małgorzata Latocha, Anna Mertas, Zenon Czuba, Stanisław Boryczka (2017). Novel Triazole Hybrids of Betulin: Synthesis and Biological Activity Profile. *Molecules*, 22: 1876.

[160] Campos Vinícius R., Paula A. Abreu, Helena C. Castro, Carlos R. Rodrigues, Alessandro K. Jordão, Vitor F. Ferreira, Maria C. B. V. de Souza, Fernanda da C. Santos, Laura A. Moura, Thaisa S. Domingos, Carla Carvalho, Eládio F. Sanchez, André L. Fuly, Anna C. Cunha (2009). Synthesis, biological, and theoretical evaluations of new 1,2,3-triazoles against the hemolytic profile of the *Lachesis muta* snake venom. *Bioorganic & Medicinal Chemistry,* 17: 7429-7434.

[161] Domingos Thaisa F. S., Laura de A. Moura, Carla Carvalho, Vinícius R. Campos, Alessandro K. Jordão, Anna C. Cunha, Vitor F. Ferreira, Maria Cecília B. V. de Souza, Eladio F. Sanchez, André L. Fuly (2013). Antivenom Effects of 1,2,3-Triazoles against *Bothrops jararaca* and *Lachesis muta* Snakes. *BioMed Research International*, Article ID 294289, 1-7.

[162] Im Haeuk, Samarendra N. Maiti, Ronald G. Micetich, Mohsen Daneshxalab, Kevin Atchison, Oludotun A. Phillips, Chieko Kunugita (1994). Synthesis and beta-lactamase inhibitory activity of 6-[(1-heteroarylthioethyl-1,2,3-triazol-4-yl)-methylene]penam sulfones. *The Journal of Antibiotics*, 47: 1030-1040.

[163] Menegatti Ricardo, Anna C. Cunha, Vítor F. Ferreira, Edna F. R. Perreira, Ahmed El-Nabawi, Amira T. Eldefrawi, Edson X. Albuquerque, Gilda Neves, Stela M. K. Rates, Carlos A. M. Fraga, Eliezer J. Barreiro (2003). Design, synthesis and pharmacological profile of novel dopamine D2 receptor ligands. *Bioorganic & Medicinal Chemistry*, 11: 4807-4813.

[164] Pålhagen Sven, Raffaele Canger, Olaf Henriksen, Johan A. van Parys, Marie-Emmanuelle Rivière, Mary Ann Karolchyke. (2001). Rufinamide: a double-blind, placebo-controlled proof of principle trial in patients with epilepsy. *Epilepsy Research*, 43: 115-124.

[165] Monceaux Christopher J., Chiho Hirata-Fukae, Polo C.-H. Lam, Maxim M. Totrov, Yasuji Matsuok, Paul R. Carliera (2011). Triazole-linked reduced amide isosteres: An approach for the fragment-based drug discovery of anti-Alzheimer s BACE1 inhibitors. *Bioorganic & Medicinal Chemistry,* 21: 3992-3996.

[166] Risho Hiroshi H., Miyoshi Yasuaki Takasu, Hirokazu Kai, Masaki Kurogochi, Shingo Arioka, Xiao-Dong Gao, Nobuaki Miura, Naoki Fujitani, Shinya Omoto, Tomokazu Yoshinaga, Tamio Fujiwara, Takeshi Noshi, Hiroko Togame, Hiroshi Takemoto, Shin-Ichiro Nishimura (2011). A strategy for neuraminidase inhibitors using mechanism-based labeling information. *Chemistry an Asian Journal,* 6: 1048-56.

[167] Kai Hirokazu, Hiroshi Hinou, Shin-Ichiro Nishimura (2012). Aglycone-focused randomization of 2-difluoromethylphenyl-type sialoside suicide substrates for neuraminidases. *Bioorganic & Medicinal Chemistry*, 20: 2739–2746.

[168] Borgati Thiago F., Rosemeire B. Alves, Róbson R. Teixeira, Rossimiriam P. de Freitas, Thays G. Perdigão, Silma F. da Silva, Aline A. dos Santos, Alberto de J. O. Bastidas (2013). Synthesis and

Phytotoxic Activity of 1,2,3-Triazole Derivatives. *Journal of the Brazilian Chemical Society*, 24: 953-961.

[169] Rastegari Arezoo, Hamid Nadri, Mohammad Mahdavi, Alireza Moradi, Seyedeh S. Mirfazli, Najmeh Edrakie, Farshad Homayouni Moghadam, Bagher Larijani, Tahmineh Akbarzadeh, Mina Saeedi (2019). Design, synthesis and anti-Alzheimer's activity of novel 1,2,3-triazolechromenone carboxamide derivatives. *Bioorganic Chemistry*, 83: 391-40.

[170] Ferreira Vitor F., David da Rocha, Fernando de C. da Silva, Patrícia G. Ferreira, Núbia A. Boechat, Jorge L. Magalhães (2013). Novel 1H-1,2,3-, 2H-1,2,3-, 1H-1,2,4- and 4H-1,2,4-triazole derivatives: a patent review (2008 - 2011). *Journal Expert Opinion on Therapeutic Patents,* 23: 319-31.

[171] Mandal Sudeep K., Dibyajyoti Saha, Vibhor K. Jain1, Bindu Jain (2010). Sythesis and antitubercular activity of some triazole derivatives of propyl gallate. International *Journal of Pharma Sciences and Research,* 1: 465-472.

[172] Chrysina Evangelia D., Éva Bokor, Kyra-Melinda Alexacou, Maria-Despoina Charavgi, George N. Oikonomakos, Spyros E. Zographos, Demetres D. Leonidas, Nikos G. Oikonomakos, László Somsák (2009). Amide-1,2,3-triazole bioisosterism: the glycogen phosphorylase case. *Tetrahedron: Asymmetry*, 20: 733-740.

[173] Bock Victoria D., Dave Speijer, Henk Hiemstra, Jan H. van Maarseveen (2007). *Organic & Biomolecular Chemistry*, 5: 971-975.

[174] Horne W. Seth, Maneesh K. Yadav, C. David Stout, M. Reza Ghadiri (2004). Heterocyclic peptide backbone modifications in an alpha-helical coiled coil. *Journal of American Chemistry Society*, 126: 15366-15367.

[175] Bezouska Karel (2002). Design, functional evaluation and biomedical applications of carbohydrate dendrimers (glycodendrimers). *Reviews in Molecular Biotechnology,* 90: 269-290.

[176] Rajeev Kharb, Prabodh C. Sharma, Mohammed S. Yar (2011). Pharmacological significance of triazole scaffold. *Journal of Enzyme Inhibition and Medicinal Chemistry*, 26: 1-21.

[177] Yan Ze-Yi, Yan-Ning Niu, Hai-Long Wei, Lu-Yong Wu, Ya-Bin Zhao, Yong-Min Liang (2006). Combining proline and 'click chemistry': a class of versatile organocatalysts for the highly diastereo- and enantioselective Michael addition in water. *Tetrahedron: Asymmetry*, 17: 3288-3293.

[178] Chandrasekhar Srivari, Togapur P. Kumar, Kothapalli Haribabu, Chada R. Reddy (2010). Hydroxyphthalimide allied triazole-pyrrolidine catalyst for asymmetric Michael additions in water. *Tetrahedron: Asymmetry*, 21: 2372-2375.

[179] Chandrasekhar Srivari, Togapur P. Kumar, Kothapalli Haribabu, Chada R. Reddy (2010). Synthesis of hybrid 1,2,3-triazolo-δ-lactams/lactones using Huisgen [3+2] cycloaddition 'click-chemistry' in water. *Tetrahedron: Asymmetry,* 21: 352-355.

[180] Zhao Ya-Bin, Lin-Wei Zhang, Lu-Yong Wu, Xing Zhong, Rong Li, Jian-Tai Ma (2008). Silica-supported pyrrolidine-triazole, an insoluble, recyclable organocatalyst for the enantioselective Michael addition of ketones to nitroalkenes. *Tetrahedron: Asymmetry*, 19: 1352-1355.

[181] Zammit Charlotte M., MartinWills (2013). Use of triazole-ring formation to attach a Ru/TsDPEN complex for asymmetric transfer hydrogenation to a soluble polymer. *Tetrahedron: Asymmetry*, 24: 844-852.

[182] Yoshida Yasushi, Shinobu Takizawa, Hiroaki Sasai (2012). Design and synthesis of spiro bis(1,2,3-triazolium) salts as chiral ionic liquids. *Tetrahedron: Asymmetry,* 23: 843-851.

[183] Raiguel Stijn, Joice Thomas, Koen Binnemans, Wim Dehaen (2018). Multi-Gram Scale Synthesis of 1,2,3-Triazolium Ionic Liquids and Assay of Their Resistance towards Bases. *European Journal of Organic Chemistry*, 35: 4850-4856.

[184] Puguan John M. C., Lilian B. Boton, Hern Kim (2018). Triazole-based ionene exhibiting tunable structure and ionic conductivity

obtained via cycloaddition reaction: A new polyelectrolyte for electrochromic devices. *Solar Energy Materials and Solar Cells*, 188: 210-218.

[185] Nguyen Manh L., Jaeduk Byun, Suwoong Kim, June W. Hyun K. Hur, Tae J. Shin, Byoung-Ki Cho (2019). Ferroelectrically Switching Helical Columnar Assembly Comprising Cisoid Conformers of a 1,2,3-Triazole-based Liquid Crystal. *Angewandte Chemie International Edition*, 58: 2749-2753.

[186] Dias Flaviana R. F., Juliana S. Novais, Talita A. do N. S. Devillart, Wanderson A. da Silva, Matheus O. Ferreira, Raquel de S. Loureiro, Vinicius R. Campos, Vitor F. Ferreira, Maria C. B. V. de Souza, Helena C. Castro, Anna C. Cunha (2018). Synthesis and antimicrobial evaluation of amino sugar-based naphthoquinones and isoquinoline-5,8-diones and their halogenated compounds. *European Journal of Medicinal Chemistry,* 156: 1-12.

[187] Bouhadir Kamal, Hala Atallah, Rana Mezher, Maamoun Fatfat, Hala Gali-Muhtasib, Jomana Elaridi (2017). Synthesis and biological assessment of novel acylhydrazone derivatives of 2-methyl-1,4-naphthoquinone. *Organic Communications,* 10: 259-272.

[188] García-Barrantes Pedro M., Guy V. Lamoureux, Alice L. Pérez, Rory N. García-Sánchez, Antonio R. Martínez, Arturo S. Feliciano (2013). Synthesis and biological evaluation of novel ferrocene-naphthoquinones as antiplasmodial agents. *European. Journal Medicinal Chemistry*, 70: 548-557.

[189] Hughes Louise M., Charlotte A. Lanteri, Michael T. O'Neil, Jacob D. Johnson, Gordon W. Gribble, Bernard L. Trumpower (2011). Design of anti-parasitic and anti-fungal hydroxy-naphthoquinones that are less susceptible to drug resistance. *Molecular and Biochemical Parasitology,* 177: 12-19.

[190] de Mello Camilly P. P., Nathália S. Sardoux, Luciana Terra, Leonardo C. Amorim, Maria D. Vargas, Gustavo B. da Silva, helena C. Castro, Viveca A. Giongo, Lucianne F. Madeira, Izabel C. N. P. Paixão (2016), Aminomethylnaphthoquinones and HSV-1: in vitro

and in silico evaluations of potential antivirals. *Antiviral Therapy*, 21: 507-515.

[191] Mandke A. D., N. R. Gonewar, R. G. Sarawadekar, A. B. Pawar (2017), Synthesis and spectral studies of 2-amino-1,4-Naphthoquinone. *International Journal of Chem Tech Research*, 10: 291-298.

[192] Pereyra Carolina E., Rafael F. Dantas, Sabrina B. Ferreira, Luciano P. Gomes, Floriano P. Silva-Jr (2019). The diverse mechanisms and anticancer potential of naphthoquinones. *Cancer Cell International*, 19: 207.

[193] Qiu Han-Yue, Peng-Fei Wang, Hong-Yan Lin, Cheng-Yi Tang, Hai-Liang Zhu, Yong-Hua Yang (2018). Naphthoquinones: A continuing source for discovery of therapeutic antineoplastic agents. *Chemical Biology & Drug Design*, 91: 681-690.

[194] Subramanian Shylaja, Márcia M. C. Ferreira, Milan Trsic (1998). A Structure-Activity Relationship Study of Lapachol and Some Derivatives of 1,4-Naphthoquinones Against Carcinosarcoma Walker 256. *Structural Chemistry*, 9: 47-57.

[195] Rao K. V., T. J. Mcbride, J. J. Oleson (1968). Recognition and Evaluation of Lapachol as an Antitumor Agent. *Cancer Research,* 28: 1952-1954.

[196] Lima Nadja M. F., Aldenir F. Santos, Zenaldo Porfírio, Marília O. F Goulart, Antônio E. G. Sant'Ana (2002). Toxicity of lapachol and isolapachol and their potassium salts against *Biomphalaria glabrata*, *Schistosoma mansoni cercariae*, *Artemia salina* and *Tilapia nilotica*. *Acta Tropica*, 83: 43-47.

[197] Lima Nadja M. F., Clariane S. Correia, Leonor L. Leon, Gérzia M. C. Machado, Maria de Fátima Madeira, Antônio E. G. Santana, Marília O. F. Goulart (2004). Antileishmanial activity of lapachol analogues. *Memória do Instituto Oswaldo Cruz*, 99(7): 757-761.

[198] Weaver, R. J; Dickins, M.; Burke, M. D. (1993). Cytochrome P450 2C9 is responsible for hydroxylation of the naphthoquinone antimalarial drug 58C80 in human liver. *Biochemical Pharmacology,* 46: 1183-1197.

[199] Canfield, C. J.; Pudney, M.; Gutteridge, W. E. (1995). Interaction of atovaquone antimalarial drugs against *Plasmodium falciparum* in vitro. *Experimental Parasitology,* 80: 373-381.

[200] Olliaro, P.; Wirth, P. D. (1997). New Targets for antimalarial drug discovery. *Journal of Pharmacy and Pharmacology,* 49: 29-33.

[201] Ball, M. D.; Bartlett, M. S.; Shaw, M.; Smith, J. W.; Nasr, M.; Meshnick, S. R. (2001). Activities and Conformational Fitting of 1,4-Naphthoquinone Derivatives and Other Cyclic 1,4-Diones Tested In Vitro against *Pneumocystis carinii. Antimicrobial Agents Chemotherapy*, 45: 1473-1479.

[202] Färnert, A.; Lindberg, J.; Gil, P.; Swedberg, G.; Berqvist, Y.; Thapar, M. M.; Lindegårdh, N.; Berezcky, S.; Björkman A. (2003). Evidence of Plasmodium falciparum malaria resistant to atovaquone and proguanil hydrochloride: case reports. *British Medical Journal*, 326: 628-629.

[203] Hashemi-Fesharki, R. Chemotherapeutic value of parvaquone and buparvaquone against *Theileria annulata* infection of cattle. *Res. Vet. Sci.,* 1991, 50, 204-207.

[204] da Silva F. C., Ferreira, V. F. (2016). Natural naphthoquinones with great importance in medicinal chemistry. *Current Organic Synthesis,* In Press.

[205] Chen J., J. Xie, Z. Jiang, B. Wang, Y. Wang, X. Hu (2011). Shikonin and its analogs inhibit cancer cell glycolysis by targeting tumor pyruvate kinase-M2. *Oncogene*, 30: 4297-4306.

[206] Fakzhenfeng Hao, Jing Qian, Jishi Yang. (2005). Shikonin induces apoptosis and inhibits migration of ovarian carcinoma cells by inhibiting the phosphorylation of Src and FAK. *Oncology Letters*, 9: 629-633.

[207] Rao, K. V.; Cullen, W. P. Streptonigrin, an antitumor substance. I. Isolation and characterization (1959-1960). *Antibiotics Annual,* 7: 950.

[208] Von Hoff D. D., Rozencweig M., Soper W. T., Helman L. J., Penta J. S., Davis H. L., Muggia F. M. (1997). Whatever happened to

NSC--? An analysis of clinical results of discontinued anticancer agents. *Cancer Treatment Reviews*, 61: 759-768.

[209] Nagaoka Hiroto, Yoshito Kishi (1981). Further Synthetic Studies on Rifamycin's. *Tetrahedron*, 37: 3873-3888.

[210] Omura Satoshi, Akira Nakagawa, Haruki Yamada, Toju Hata, Akio Furusaki, Tokunosuke Watanaee (1973). Structures and Biological Properties of Kinamycin A, B. C and D. *Chemical and Pharmaceutical Bulletin*, 21: 931-940.

[211] Brockman William W., William A. Carter, Ernest C. Borden (1971). Streptovaricins inhibit Focus Formation by MSV (MLV) Complex. *Nature New Biology*, 232: 212-214.

[212] Jiang Yuanyuan, Li Zhang, HP V. Rupasinghe (2016). The anticancer properties of phytochemical extracts from *Salvia* plants. *Botanics: Targets and Therapy,* 6: 25–44.

[213] Luo Hua, Chi Teng Vong, Hanbin Chen, Yan Gao, Peng Lyu, Ling Qiu, Mingming Zhao, Qiao Liu, Zehua Cheng, Jian Zou, Peifen Yao, Caifang Gao, Jinchao Wei, Carolina Oi Lam Ung, Shengpeng Wang, Zhangfeng Zhong, Yitao Wang (2019). Naturally occurring anti-cancer compounds: shining from Chinese herbal medicine. *Chinese Medicine,* 14: 1-58.

[214] Zhu Zhongcheng (2018). Miltirone-induced apoptosis in cisplatin-resistant lung cancer cells through upregulation of p53 signaling pathways. *Oncology Letters,* 15: 8841-8846.

[215] Wu Xiao, Benjamin Chantemargue, Florent Di. Meo, Claudie Bourgaux, David Chapron, Patrick Trouillas, Veronique Rosilio (2019). Deciphering the Peculiar Behavior of β-Lapachone in Lipid Monolayers and Bilayers. *Langmuir,* 35: 14603-14615.

[216] Fieser Louis F. (1937). 1,2-naphthoquinone. *Organic Syntheses,* 17: 68.

[217] Sim Jaeuk, Hyeju Jo, Mayavan Viki, Minho Choi, Jin-Ah Jung, Heesoon Lee, Jae-Kyung Jung (2018). Rapid, Operationally Simple, and Metal-free NBS Mediated Onepot Synthesis of 1,2-Naphthoquinone from 2-Naphthol. *Advanced Synthesis & Catalysis,* 360: 852-858.

[218] Shukla Gaurav, Abhijeet Srivastava, Dhananjay Yadav, Maya S. Singh (2018). Copper-Catalyzed One-Pot Cross-Dehydrogenative Thienannulation: Chemoselective Access to Naphtho[2,1-b] thiophene-4,5-diones and Subsequent Transformation to Benzo[a]thieno[3,2-c]phenazines. *The Journal of Organic Chemistry,* 83: 2173-2181.

[219] Fieser Louis F. (1925). 1,4-naphthoquinone. *Organic Syntheses*, 5: 79.

[220] Braude E. A., J. S. Fawcett (1953). 1,4-naphthoquinone. *Organic Syntheses,* 33: 50.

[221] Jordão Alessandro K., Maria D. Vargas, Angelo C. Pinto, Fernando de C. da Silva, Vitor F. Ferreira (2015). Lawsone in organic synthesis. *RSC Advances*, 5: 67909-67943.

[222] Gogin Leonid L., Elena G. Zhizhina, Zinaida P. Pai (2019). One-Pot Process of Naphthoquinones Synthesis from Hydroquinone in the Presence of Solutions of Mo-V-P Heteropolyacids as Bifunctional Catalysts. *Modern Research in Catalysis,* 8: 1-9.

[223] Jardim Guilherme A. M., Eufrânio N. da Silva Junior, John F. Bower (2016). Overcoming naphthoquinone deactivation: rhodium-catalyzed C-5 selective C–H iodination as a gateway to functionalized derivatives. *Chemical Science,* 7: 3780.

[224] Jardim Guilherme A. M., Willian X. C. Oliveira, Rossimiriam P. de Freitas, Rubem F. S. Menna-Barreto, Thaissa L. Silva, Marília O. F. Goulart, Eufrânio N. da Silva Júnior (2018). Direct sequential C-H iodination/organoylthiolation for the benzenoid A-ring modification of quinonoid deactivated systems: a new protocol for potent trypanocidal quinones. *Organic and Biomolecular Chemistry,* 16: 1686.

[225] Jardim Guilherme A. M., Ícaro A. O. Bozzi, Willian X. C. Oliveira, Camila Mesquita-Rodrigues, Rubem F. S. Menna-Barreto, Ramar A. Kumar, Edmond Gravel, Eric Doris, Antonio L. Braga, Eufrânio N. da Silva Júnior (2019). Copper complexes and carbon nanotube–copper ferrite-catalyzed benzenoid A-ring selenation of quinones: an

efficient method for the synthesis of trypanocidal agents. *New Journal of Chemistry,* 43**:** 13751-13763.

[226] Kurban Semih, Nahide G. Deniz, Cigdem Sayil, Mustafa Ozyurek, Kubilay Guclu, Maryna Stasevych, Viktor Zvarych, Olena Komarovska-Porokhnyavet, Volodymyr Novikov (2019). Synthesis, Antimicrobial Properties, and Inhibition of Catalase Activity of 1,4-Naphtho- and Benzoquinone Derivatives Containing N-, S-, O-Substituted. *Heteroatom Chemistry,* 2019: 1-12.

[227] Matsui Takuya, Naohito Yamada, Hideyuki Kuno, Robert A. Kanaly (2019). Formation of Bulky DNA Adducts by Non-Enzymatic Production of 1,2-Naphthoquinone-Epoxide from 1,2-Naphthoquinone under Physiological Conditions. *Chemical Research in Toxicology,* 32: 1760-1771.

[228] Chipoline Ingrid C., Vitor F. Ferreira, Fernando de C. da Silva (2015). One-Step Synthesis of 1H-1,2,3-Triazol-1-Ylmethyl-2,3-Dihydronaphtho[1,2-b]-furan-4,5-Diones. *Current Organic Synthesis,* 12: 565-569.

[229] Tsanakopoulou Maria, Erifili Tsovaltzi, Marina A. Tzani, Periklis Selevos, Elizabeth Malamidou-Xenikaki, Evangelos G. Bakalbassisb, Luis R. Domingo (2019). Site-selectivity control in hetero-Diels–Alder reactions of methylidene derivatives of lawsone through modification of the reactive carbonyl group: an experimental and theoretical study. *Organic & Biomolecular Chemistry,* 17(3): 692.

[230] Shangguan Yu, Fazhou Yang, Hao Deng, Hao Liu, Ziyan Liu, Wanyue Zhuang, Chenxi Qiao, Aizheng Wang, Yumei Xiao, Cheng Zhang (2019). Copper-Catalyzed Three-Component Difunctionalization of Aromatic Alkenes with 2-Amino-1,4-naphthoquinones and α-Bromocarboxylates. *The Journal of Organic Chemistry,* 84: 10649-10657.

[231] Naturale Guillaume, Marc Lamblin, Claude Commandeur, François-Xavier Felpin, Jean Dessolin (2012). Direct C–H Alkylation of Naphthoquinones with Amino Acids Through a Revisited Kochi–Anderson Radical Decarboxylation: Trends in Reactivity and

Applications. *European Journal of Organic Chemistry,* 29: 5774-5788.

[232] Ferreira Patricia G., Luana P. Borba-Santos, Leticia L. Noronha, Caroline D. Nicoletti, Marcella de Sá H. Queiroz, Fernando C. da Silva, Sônia R., Débora O. Futuro, Vitor F. Ferreira (2019). Synthesis, Stability Studies, and Antifungal Evaluation of Substituted α- and β-2,3-Dihydrofuranaphthoquinones against *Sporothrix brasiliensis* and *Sporothrix schenckii*. *Molecules,* 24: E930.

[233] Souza Daniel M., Daniela P. Lage, Débora V. C. Mendonça, Eduardo A. F. Coelho, Flaviano M. Ottoni, Grasiele de Sousa V. Tavares, Mariana C. Duarte, Ricardo J. Alves (2017). *Composições farmacêuticas leishmanicidas contendo fucosídeo peracetilado derivado da lausona e uso.* BR102017023735. [*Pharmaceutical leishmanicidal compositions containing peracetylated fucoside derived from lausone and use.*]

[234] Meimei Guo, He Jianming, Gao Yue (2019). *Application and its medicine preparation of the naphthoquinones glycosides compound in senile dementia*. CN110025624.

[235] Qinge Ma, Wei Rongrui (2018). *Naphthoquinone compounds and their preparation and application in preparing hypoglycemic products.* CN108503653.

[236] Eufrânio N. da Silva Jr.; Rubem F.S.Menna-Barreto; Maria do Carmo F.R.Pinto; Raphael S.F.Silva; Daniel V. Teixeira; Maria Cecília B.V. de Souza; Carlos Alberto De Simone; Solange L. De Castro; Vitor F. Ferreira; Antônio V. Pinto (2008). Naphthoquinoidal [1,2,3]-triazole, a new structural moiety active against Trypanosoma cruzi. *European Journal of Medicinal Chemistry*, 43: 1774-1780.

[237] Eufrânio N. da Silva Júnior; Maria Aline B. F. de Moura; Antonio V. Pinto; Maria do Carmo F. R. Pinto; Maria Cecília B. V. de SouzaI; Ana J. Araújo; Claudia Pessoa; Letícia V. Costa-Lotufo; Raquel C. Montenegro; Manoel Odorico de Moraes; Vitor F. Ferreira; Marilia O. F. Goulart (2009). Cytotoxic, Trypanocidal

Activities and Physicochemical Parameters of nor-β-Lapachone-based 1,2,3-Triazoles. *Journal of the Brazilian Chemical Society*, 20: 635-643.

[238] Guimarães Tiago T., Maria C. F.R. Pinto, Juliane S. Lanza; Maria N. Melo, Rubens L.do Monte-Neto, Isadora M.M.de Melo, Emilay B.T. Diogo, Vitor F. Ferreira, Celso A. Camara, Wagner O. Valença, Ronaldo N.de Oliveira, Frédéric Frézard, Eufrânio N.da Silva Júnior (2013). Potent naphthoquinones against antimony-sensitive and -resistant Leishmania parasites: Synthesis of novel α- and nor-α-lapachone-based 1,2,3-triazoles by copper-catalyzed azide–alkyne cycloaddition. *European Journal of Medicinal Chemistry*, 63: 523-530.

[239] Bahia Samara B. B. B., Wallace J. Reis, Guilherme A. M. Jardim, Francielly T. Souto, Carlos A. de Simone, Claudia C. Gatto, Rubem F. S. Menna-Barreto, Solange L. de Castro, Bruno C. Cavalcanti, Claudia Pessoa, Maria H. Araujo, Eufrânio N. da Silva Júnior (2016). Molecular hybridization as a powerful tool towards multitarget quinoidal systems: synthesis, trypanocidal and antitumor activities of naphthoquinone-based 5-iodo-1,4-disubstituted-, 1,4- and 1,5-disubstituted-1,2,3-triazoles. *Medicinal Chemistry Communications*, 7: 1555-1563.

[240] da Silva Júnior E. N., I. M. M. de Melo, E. B. T. Diogo, V. A. Costa; J. D. de Souza Filho; W. O. Valença; C. A. Camara; R. N. de Oliveira; A. S. de Araujo; F. S. Emery; M. R. dos Santos; C. A. de Simone; R. F. S. Menna-Barreto; S. L. de Castro (2012). On the search for potential anti-Trypanosoma cruzi drugs: Synthesis and biological evaluation by click chemistry reactions. *European Journal of Medicinal Chemistry,* 52: 304-312.

[241] Pacheco P. A. F., R. M. S. Galvão, A. F. M. Faria, N. l. Von Ranke, M. S. Rangela, T. M. Ribeiro, M. l. Bello, C. R. Rodrigues, V. F. Ferreira, D. R. da Rocha, R. X. Faria (2019). 8-Hydroxy-2-(1H-1,2,3-triazol-1-yl)-1,4-naphtoquinone derivatives inhibited P2X7 Receptor-Induced dye uptake into murine Macrophages. *Bioorganic & Medicinal Chemistry*, 27: 1449-1455.

[242] da Cruz Eduardo H. G., Caio M. B. Hussene, Gleiston G. Dias, Emilay B. T. Diogo, Isadora M. M. de Melo, Bernardo L. Rodrigues, Mauro G. da Silva, Wagner O. Valença, Celso A. Camara, Ronaldo N. de Oliveira, Yen G. de Paiva, Marilia O. F. Goulart, Bruno C. Cavalcanti, Claudia Pessoa, Eufrânio N.da Silva Júnior (2014). 1,2,3-Triazole-, arylamino- and thio-substituted 1,4-naphthoquinones: Potent antitumor activity, electrochemical aspects, and bioisosteric replacement of C-ring-modified lapachones, *Bioorganic & Medicinal Chemistry*, 22: 1608-1619.

[243] Gholampour Maryam, Sara Ranjbar, Najmeh Edraki, Maryam Mohabbati, Omidreza Firuzi, Mehdi Khoshneviszadeh (2019). Click chemistry-assisted synthesis of novel aminonaphthoquinone-1,2,3-triazole hybrids and investigation of their cytotoxicity and cancer cell cycle alterations. *Bioorganic Chemistry*, 88: 102967.

[244] Brandão Geraldo Célio, Franciele C.Rocha, Missias Lucas, Miquéias Arantes, Luciana Ferreira Soares, Kuldeep K.Roy, Robert J. Doerksen, Alaide Braga de Oliveira, Guilherme Rocha Pereira (2018). Antimalarial naphthoquinones. Synthesis via click chemistry, in vitro activity, docking to PfDHODH and SAR of lapachol-based compounds, R of lapachol-based compounds. *European Journal of Medicinal Chemistry,* 145: 191-205.

[245] Oramas-Royo Sandra, Priscila López-Rojas, Ángel Amesty, David Gutiérrez, Ninoska Flores, Patricia Martín-Rodríguez, Leandro Fernández-Pérez, Ana Estévez-Braun (2019). Synthesis and Antiplasmodial Activity of 1,2,3-Triazole-Naphthoquinone Conjugates. *Molecules*, 24: 3917.

[246] Aneja Babita, Mudsser Azam, Shadab Alam, Ahmad Perwez, Ronan Maguire, Umesh Yadava, Kevin Kavanagh, Constantin G. Daniliuc, M. M. A. Rizvi, Qazi M. R. Haq, Mohammad Abid (2018). Natural Product-Based 1,2,3-Triazole/Sulfonate Analogues as Potential Chemotherapeutic Agents for Bacterial Infections. *ACS Omega*, 3: 6912-6930.

[247] Prasad Chakka V., Vadithe L. Nayak, Sistla Ramakrishna, Uppuluri V. Mallavadhani (2017). Novel menadione hybrids: Synthesis,

anticancer activity, and cell-based studies. *Chemical Biology & Drug Design*, 91(1): 220-233.

[248] Prasad Chakka V., Vadithe L. Nayak, Sistla Ramakrishna, Uppuluri V. Mallavadhani (2017). Novel menadione hybrids: Synthesis, anticancer activity, and cell-based studies. *Chemical Biological & drug design*, 91: 220-233.

[249] Braga H. C., Stefani H. A., Paixão M. W., Santos F. W., Lüdtke D. S. (2019). Synthesis of 50-selenoxylofuranosides. *Tetrahedron*, 66: 3441-3446.

[250] Alberto Eduardo E., Luana L. Rossato, Sydney Hartz Alves, Diego Alves, Antonio L. Braga (2011). Imidazolium containing selenium: synthesis and antimicrobial activity. *Organic & Biomolecular activity*, 9: 1001-1003.

[251] Jardim Guilherme A. M., Daisy J. B. Lima, Wagner O. Valença, Daisy J. B. Lima, Bruno C. Cavalcanti, Claudia Pessoa, Jamal Rafique, Antonio L. Braga, Claus Jacob, Eufrânio N. Da Silva Júnior, Eduardo H. G. Da Cruz (2018). Synthesis of Selenium-Quinone Hybrid Compounds with Potential Antitumor Activity via Rh-Catalyzed C-H Bond Activation and Click Reactions. *Molecules*, 23: 83-98.

[252] Costa Dora C. S., Gabriella S. de Almeida, Vitor W. Rabelo, Lucio M. Cabral, Plínio C. Sathler, Paula A. Abreu, Vitor. F. Ferreira, Luiz C. R. P. da Silva, Fernando de C. da Silva (2018). Synthesis and evaluation of the cytotoxic activity of Furanaphthoquinones tethered to 1H-1,2,3-triazoles in Caco-2, Calu-3, MDA-MB231 cells. *European Journal of Medicinal Chemistry*, 156: 524-533.

[253] Chipoline Ingrid C., Evelyne Alves, Paola Branco, Leticia V. Costa-Lotufo, Vitor F. Ferreira, Fernando C. Da Silva (2018). Synthesis and Cytotoxic Evaluation of 1H-1,2,3-Triazol-1- ylmethyl-2,3-dihydronaphtho[1,2-b]furan-4,5-diones. *Anais da Academia Brasileira de Ciências*, 90: 1027-1033.

[254] Paula F. Carneiro, Maria do Carmo F. R. Pinto, Roberta K. F. Marra, Vinícius R. Campos, Jackson Antônio L. C. Resende, Maicon Delarmelina, José Walkimar M. Carneiro, Emerson S. Lima,

Fernando de C. da Silva, Vitor F. Ferreira (2016). Insight into and Computational Studies of the Selective Synthesis of 6H-Dibenzo[b,h]xanthenes. *The Journal of Organic Chemistry*, 81: 5525-5537.

[255] Bortolot Carolina S., Luana da S.M. Forezi, Roberta K.F. Marra, Marcelo I. .P. Reis, Bárbara V. F. e Sá, Ricardo I. Filho, Zeinab Ghasemishahrestani, Mauro Sola-Penna, Patricia Zancan, Vitor F. Ferreira, Fernando de C. da Silva (2019). Design, Synthesis and Biological Evaluation of 1H-1,2,3-Triazole-Linked-1H-Dibenzo [b,h]xanthenes as Inductors of ROS-Mediated Apoptosis in the Breast Cancer Cell Line MCF-7. *Medicinal Chemistry*, 15: 119-129.

[256] Branco Jessica R., Vanessa G. Oliveira, Amanda M. Esteves, Ingrid C. Chipoline, Miriam F.O. Lima, Fernanda C.S. Boechat, Fernando C. da Silva, Vítor F. Ferreira, Mauro Sola-Penna, Maria C.B.V. de Souza, and Patricia Zancan (2018). A Novel Naphthotriazolyl-4-oxoquinoline Derivative that Selectively Controls Breast Cancer Cells Survival Through the Induction of Apoptosis. *Current Topics in Medicinal Chemistry*, 18: 1465-1474.

BIOGRAPHICAL SKETCH

Vitor Francisco Ferreira

Affiliation:
Universidade Federal Fluminense, Faculdade de Farmácia, Departamento de Tecnologia Farmacêutica, 24241-000, Niterói, Brazil

Education: PhD in Chemistry at University of California San Diego

Research and Professional Experience: My scientific skill involves the areas of Organic Chemistry and Medicinal Chemistry, with emphasis on Organic Synthesis focused on asymmetric methodologies and bioactive molecules in the areas of cancer, tuberculosis, Chagas disease. His works in chemistry mainly involve the following topics: carbohydrates (monosaccharides), nucleosides, diazocompostos, heterocycles and naphthoquinones. His research looks at new compounds and new ways of formulating compounds used to treat cancer, AIDS and neglected infectious diseases. A significant number of his publications deals with the synthesis and biological evaluation of new amphibian, antitubercular, anticancer, antiviral, trypanocidal and some publications also concerns pro-drugs. The present invention relates to a method for the preparation of anti-inflammatory and antigenic microorganisms, virus, cancer, HIV, fungi, bacteria, as well as diabetes treatment and snake bites. His work has always been focused on the synthesis of analogues of α-xyloidones, α and β-natural lapachones that present considerable activities against S. mansoni and T. cruzi, the latter being the microorganism that causes Chagas Disease and also showed high antitumor activity, especially in solid tumors. It worked with nickel-based catalysts by exploring a type of selective coupling of the Grignard reagent with enol ethers, catalyzed by nickel (0), in the sesquiterpene synthesis of the cadinan group and some important acyclic pheromones. A particularly interesting work was 1,5-naphthalenediol oxidation to obtain the natural juglone naphthoquinone, which has allelopathic activity, with the NaIO4 / DMF system which has been used by many research groups as a starting material for many syntheses. He used this compound for the preparation of lapachone substituted on the aromatic ring. In another line of research studied the use of sucrose as raw material for the preparation of products with potential technological applications in several areas of Sucroquímica that have become publications and patent. He developed a method of hydrolysis of Sucrose with carbon dioxide that gave him a patent granted by INPI (Brazil) and third place in the Contest of the National Inventor. He was the coordinator of the project, with federal resources, to synthesize new nucleosides of non-conventional nitrogen bases, and to test them against viruses and bacteria. This Nucleoside Research Group was the first to establish itself in Brazil. The investigations involved structural modifications due to alterations in the carbohydrate, in the nitrogenous base or in the type or position of the glycosidic bond. He created the laboratory of organic synthesis and natural products with biological applications. He was also one of the creators of the Nuclear Magnetic Resonance Center and the Regional Laboratory of X-Ray Diffractometry. He was one of the founders of the Master's and Doctoral courses in Organic Chemistry at the Institute of Chemistry and the Master's and Doctorate courses in Applied Health Products at the Faculty of Pharmacy. All of these facilities were created within the Universidade Federal Fluminense. These facilities and courses were responsible for training hundreds of masters and doctors. The creation of postgraduate programs within a public institution having to overcome all the difficulties of lack of resources, personnel and infrastructure is a task and legacy that is very noble and relevant to science. He was the Secretary General of the 4th Brazilian Meeting in Organic Synthesis, which was notable for bringing eight first-class foreign lecturers together with our PG Teachers and Students. He was the author along with three other textbook teachers, "Acids and Bases in Organic Chemistry," who sold more than thousand copies in Brazil and other Portuguese-speaking countries, and can still be found in bookstores. In response to the invitation of Profs. Wim Dehaen and Vasiliy A. Bakulev wrote the chapter "Topics in Heterocyclic Chemistry" which was published by the publisher Springer-Verlag Berlin; more specifically, the chapter was concerned with the biological properties of the triazoles "Biological properties of 1H-1,2,3- and 2H-1,2,3-triazoles." During the annual ACS meeting in New Orleans, he was invited by the president of the American Chemistry Society (ACS) Marinda Li Wu to attend a panel discussion on "Vision 2025: How to Succeed in the Global Chemistry Enterprise." After the round table was invited to write a chapter of a book that would be edited by the ACS Symposium Series: "The Brazilian Chemical Society (SBQ) and the Global Chemistry Enterprise: Building a Sustainable Development Strategy."

My professional appointments also highlight intense political activities, because I believe that there is no economically developed country that does not also have a large chemical production and research in frontier areas of knowledge. The production of chemical inputs generates thousands of jobs and is also responsible for a significant part of GDP.

Scientific and technological knowledge highlights the most advanced and economically stronger countries, especially when they transform knowledge into innovation.

Professional Appointments: Professor Vitor Francisco Ferreira was born in Rio de Janeiro/Brazil in 1953. He was Researcher fellow at the National Institute of Technology, 1985-1990; Professor of Organic Chemistry, Fluminense Federal University, Rio de Janeiro since 1986 Brazil; Full Professor of Organic Chemistry, Fluminense Federal University, Rio de Janeiro since 1993-1997; Full Professor Pharmaceutical Technology, Fluminense Federal University, Rio de Janeiro since 1997 Brazil; CNPq researcher since 1988 and currently as a researcher level 1A; Master in Natural Products Chemistry, UFRJ, IPPN, 1980, Brazil; Master in Chemistry, University of California San Diego, 1982, USA; PhD in Chemistry, University of California San Diego, 1984, USA in the group of Prof. Ernest Wenkert on three research fronts, namely: alkaloid synthesis, chemistry of α-diazocarbonyl compounds and organometallic compounds derived from nickel. My PhD thesis work was chosen in the diazo-α-carbonylated area, or more accurately the "Diazomalonaldehyde Chemistry Study." In continuity to his scientific career he went to postdoctoral in Organic Chemistry at University of Oklahoma (1997, USA); Senior Scientist of Research Foundation of Rio de Janeiro (since its first edition); Vice-Director of the Chemistry Institute of the Fluminense Federal University 1991-1995, Brazil; Sub-Coordinator of the Graduate Course in Organic Chemistry, Fluminense Federal University, 2002-2003; Coordinator of Research and graduate of the Dean of Research and Graduate of Fluminense Federal University, 1998-2000, Brazil; Member of the Brazilian Academy of Science since 2007; Member of the National Order of Merit Scientific-class Comendador given by the President of the Republic (2009); Former Editor of the Journal Química Nova of the Brazilian Chemistry Society; Editor of Revista Virtual de Química of the Brazilian Chemistry Society, Brazil (current); Vice President of the Brazilian Chemistry Society between 1994-1996 and 2010-2012; President of the Brazilian Chemistry Society, between 2012-2014; Fellow of the

Royal Society of Chemistry (2014), England; Dean of Research and Graduate officer of Fluminense Federal University, 2017-2018.

In 2006 he coordinated the international agreement with Portugal. with Prof. José Abrunheiro da Silva Cavaleiro from the University of Aveiro (UA) in "Porphyrins and saccharide diazo-esters: synthesis of new tetrapyrrole macrocycles with medicinal application potentialities." This agreement lasted for three years and several students went to Portugal to do doctorate - sandwich or postdoctoral. Many papers were published in the first two years and the agreement had its third year renewed in the face of the results obtained. In 2011, another Brazilian-Portuguese Scientific Exchange Program "Development of Methodology for the preparation of New Porphyrin Derivatives" was carried out with the aim of developing studies leading to new methodologies for the synthesis of compounds with potential, in view of the structural characteristics present, of medicinal application and / or agrochemicals.

He is the author of 315 scientific articles published, 10 patents, 4 books, 7 book chapters, 47 editorial of journals, more than 100 poster presented in several congress, several national and international conferences, editor of the journal. He advised 42 master dissertations, 28 PhD thesis, 50 undergraduate student and supervised 10 post-graduated fellows. He has 7618 citations, index h of 44 and index of the last 10 years (i10) of 192.

Honors: Prof. Vitor F. Ferreira received several medal and prizes. The first one was during the third year of the doctoral course at University of California San Diego (1983) named "Hart Memorial Award," dedicated to the best foreign graduate student. After then he received several other, such as, the Walter Baptista Mors Medal from Brazilian Chemical Society-Rio (2013), Lavoisier Medal from Regional Chemistry Council (CRQ) Third Region (2013). He was also honored by the Portuguese Chemistry Society in recognition of his valuable contributions in favor of closer scientific relations between Brazil and Portugal. At the end of 2018 he was awarded with the Ângelo da Cunha Pinto Medal, the highest honor awarded by the Institute of Chemistry of the Federal University of Rio de Janeiro, which aims to annually award a distinguished personality in the area of exact sciences, health sciences or humanities . The selection considered contributions for scientific development, attested by publications, public agency and S & T development, effective participation in public policies for the advancement of S & T, international insertion, among other relevant points. In May 2019, Professor Vitor Ferreira lectured at the opening conference of the 42nd Annual Meeting of the SBQ (from May 27 to 30 in Joinville) and will receive the Simão Mathias Medal, the highest SBQ award. The Simão Mathias Medal was instituted by the SBQ in 1997 to recognize personalities who have made significant contributions to Chemistry in Brazil and the SBQ.

Publications from the Last 3 Years (2016-2018):

[1] Branco, J. R.; Oliveira, V. G.; Esteves, A. M.; Chipoline, I. D. C.; Lima, M. F. O.; Boechat, F. C. S.; Da Silva, F. C.; Ferreira, V. F.; Sola-Penna, M.; Souza, M. C. B. V.; Zancan, P. 2018. A Novel Naphthotriazolyl-4-oxoquinoline Derivative that Selectively Controls Breast Cancer Cells Survival Through the Induction of Apoptosis, In *Current Topics in Medicinal Chemistry,* v.18, 1465-1474.

[2] Filho, R. I.; Gonzaga, D. T. G.; Demaria, T. M.; Leandro, J. G. B.; Costa, Dora C. S.; Ferreira, V. F.; Sola-Penna, Mauro; De C. Da Silva, F. C.; Zancan, Patricia. 2018. A Novel Triazole Derivative Drug Presenting In Vitro and In Vivo Anticancer Properties, In *Current Topics in Medicinal Chemistry,* v.18, 1483-1493.

[3] Forezi, L. S. M.; Cardoso, M. F. C.; Gonzaga, D. T. G.; De C. Da Silva, F. C.; Ferreira, V. F. 2018. Alternative Routes to the Click Method for the Synthesis of 1,2,3-Triazoles, In *Current Topics in Medicinal Chemistry,* v.18, 1428-1453.

[4] Novais, J. S.; Moreira, C. S.; Silva, A. C. J. A.; Loureiro, R. S.; Sá Figueiredo, A. M.; Ferreira, V. F.; Castro, H. C.; Da Rocha, D. R. 2018. Antibacterial naphthoquinone derivatives targeting resistant strain Gram-negative bacteria in biofilms, In *Microbial Pathogenesis,* v.118, 105-114.

[5] Simões, M.; G., Daniel; C., Cardoso, M. F. C.; Forezi, L. M. M.; Gomes, A.; Da Silva, F. C.; Ferreira, V. F.; Neves, M.; Cavaleiro, J. A. S. 2018. Carbene Transfer Reactions Catalysed by Dyes of the Metalloporphyrin Group, In *Molecules,* v.23, 792.

[6] Gomes, A. T. P. C.; Forezi, L. S. M.; Simões, M. M. Q.; Gonzaga, D. T. G.; Cardoso, M. F. C.; Da Silva, F. C.; Neves, M. G. P. M. S.; Ferreira, V. F.; Cavaleiro, J. A. S. 2018. Carbene-Type Species in the Functionalization of Porphyrin Derivatives, In *Synthesis*-(Stuttgart), v.50, 2678-2692.

[7] Da Silva, F. C.; Ferreira, V. F. 2018. Changes in the Graduate Evaluation Proceding, In *Revista Virtual de Química,* v.10, 210-210.

[8] Da Silva, F. C.; Ferreira, V. F.; Forezi, L. S. M. 2018. Editorial: Biological Profiles of Coumarin Scaffold - Part 2, In *Current Topics in Medicinal Chemistry*, v.18, 99-100.

[9] Lara, L. S.; Moreira, C. S.; Calvet, C. M.; Lechuga, G. C.; Souza, R. S.; Saulo C. Bourguignon; Ferreira, V. F.; Rocha, D. R.; Pereira, M. C. S. 2018. Efficacy of 2-hydroxy-3-phenylsulfanylmethyl-[1,4]-naphthoquinone derivatives against different Trypanosoma cruzi discrete type units: Identification of a promising hit compound, In *European Journal of Medicinal Chemistry,* v.144, 572-581.

[10] Lopes, S. M. M.; Novais, J. S.; Costa, D. C. S.; Castro, H. C.; Figueiredo, A. M. S.; Ferreira, V. F.; Pinho E .Melo, T. M. V. D.; Da Silva, F. C. 2018. Hetero-Diels-Alder reactions of novel 3-triazolyl-nitrosoalkenes as an approach to functionalized 1,2,3-triazoles with antibacterial profile, In *European Journal of Medicinal Chemistry,* v.143, 1010-1020.

[11] Gonzaga, D. T. G.; Souza, T. M. L.; Andrade, V. M. M.; Ferreira, V. F.; Da Silva, F. C. 2018. Identification of 1-Aryl-1H-1,2,3-triazoles as Potential New Antiretroviral Agents, In *Medicinal Chemistry,* v.14, 242-248.

[12] Zorzanelli, B. C; De Queiroz, L. N; Santos, R. Ma; Menezes, L. M; Gomes, F. C.; Ferreira, V. F.; Da Silva, F. C.; Robbs, B. K. 2018. Potential cytotoxic and selective effect of new benzo[b]xanthenes

against oral squamous cell carcinoma, In *Future Medicinal Chemistry,* v.10, 1141-1157.

[13] Faria, R. X.; Gonzaga, D. T. G.; Pacheco, P. A. F.; Souza, A. L. A.; Ferreira, V. F.; Da Silva, F. C. 2018. Searching for new drugs for Chagas diseases: triazole analogs display high in vitro activity against Trypanosoma cruzi and low toxicity toward mammalian cells, In *Journal of Bioenergetics and Biomembranes,* v.50, 81-91.

[14] Dias, F. R. F.; Novais, J. S.; Devillart, T. A. N. S.; Da Silva, W. A.; Ferreira, M. O.; Loureiro, R. S.; Campos, V. R.; Ferreira, V. F.; De Souza, M. C. B. V.; Castro, H. C.; Cunha, A. C. 2018. Synthesis and antimicrobial evaluation of amino sugar-based naphthoquinones and isoquinoline-5,8-diones and their halogenated compounds, In *European Journal of Medicinal Chemistry,* v.156, 1-12.

[15] Silva, W. A.; Silva, L. C. R. P.; Campos, V. R.; Souza, M. C. B. V.; Ferreira, V. F.; Santos, A. C. P. B.; Sathler, P. C.; Almeida, G. S.; Dias, F. R. F; Cabral, L. M.; Azeredo, R. B. V; Cunha, A.C. 2018. Synthesis and antitumor evaluation of hybrids of 5,8-dioxo-5,8-dihydroisoquinoline-4-carboxylates and carbohydrates, In *Future Medicinal Chemistry,* v.10, 527-540.

[16] Forezi, L. S. M.; Froes, T. Q.; Cardoso, M. F. C.; De Oliveira, M. C. A.; Nicastro, G. G.; Baldini, R. L.; Costa, D. C. S.; Ferreira, V. F.; Castilho, M. S.; Da Silva, F. C. 2018. Synthesis and Biological Evaluation of Coumarins Derivatives as Potential Inhibitors of the Production of Pseudomonas aeruginosa Virulence Factor Pyocyanin, In *Current Topics in Medicinal Chemistry,* v.18, 149-156.

[17] Da Silva, F. C.; Ferreira, V. F.; Forezi, L. S. M. 2018. Synthesis and Biological Profiles of 1,2,3-Triazole Scaffold, In *Current Topics in Medicinal Chemistry*, v.18, 1426-1427.

[18] Chipoline, I. C.; Alves, E.; Branco, P.; Costa-Lotufo, L. V.; Ferreira, V. F.; Silva, F. C. 2018. Synthesis and Cytotoxic Evaluation of 1H-1,2,3-Triazol-1-ylmethyl-2,3-dihydronaphtho[1,2-b]furan-4,5-diones, In *Anais da Academia Brasileira de Ciências,* v.90, 1027-1033.

[19] Costa, D. C. S.; De Almeida, G. S.; Rabelo, V. W.; Cabral, L. M.; Sathler, P. C.; Alvarez Abreu, P.; Ferreira, V. F.; Da Silva, C. R. P.

L.; Da Silva, F. C. 2018. Synthesis and evaluation of the cytotoxic activity of Furanaphthoquinones tethered to 1H-1,2,3-triazoles in Caco-2, Calu-3, MDA-MB231 cells, In *European Journal of Medicinal Chemistry,* v.156, 524-533.

[20] Silva, J. L.; Cino, E. A.; Soares, I. N.; Ferreira, V. F.; De Oliveira, A. P. G. 2018. Targeting the Prion-like Aggregation of Mutant p53 to Combat Cancer, In *Accounts of Chemical Research,* v.51, 181-190.

[21] Nunes, D. V. Q.; Costa, C. A.; De Bem, G. F.; Cordeiro, V. S. C.; Santos, I. B.; Carvalho, L. C. R. M.; Jordão, A. K.; Cunha, A. C.; Ferreira, V. F.; Moura, R. S; Resende, A. C.; Ognibene, D. T. 2018. Tempol, a superoxide dismutase-mimetic drug, prevents chronic ischemic renal injury in two-kidney, one-clip hypertensive rats, In *Clinical and Experimental Hypertension,* 40(8):721-729.

[22] Futuro, D. O.; Ferreira, P. G.; Nicoletti, C. D.; Borba-Santos, L. P.; Da Silva, F. C.; Rozental, S.; Ferreira, V. F. 2018. The Antifungal Activity of Naphthoquinones: An Integrative Review, In *Anais da Academia Brasileira de Ciências,* v.90, 1187-1214.

[23] Faria, R. X.; Oliveira, F. H.; Salles, J. P.; Oliveira, A. S.; Von Ranke, N. L.; Bello, M. L.; Rodrigues, C. R.; Castro, Helena Carla; Louvis, A. R.; Martins, D. L.; Ferreira, V. F. 2018. 1,4-Naphthoquinones potently inhibiting P2X7 receptor activity, In *European Journal of Medicinal Chemistry,* 143:1361-1372.

[24] Costa, D. C. S.; Forezi, L. S. M.; Cardoso, M. F. C.; Ribeiro, R. C. B.; C. Pinto, A. C.; B.; Da Silva, F. C. 2017. A Compendium of Tyrosine-kinase Inhibitors: Powerful and Efficient Drugs against Cancer, In *Revista Virtual de Química,* v.9, 974-1064.

[25] Dantas E. D., De Souza F. J. J., Nogueira W. N. L., Silva C. C., De Azevedo P. H. A, Soares Aragão C. F., Almeida P. D. O., Cardoso M. F. D. C., Da Silva F. C., De Azevedo E. P., Guimarães, E. B., Lima E. S., Ferreira V. F., Lima A. N. 2017. Characterization and Trypanocidal Activity of a Novel Pyranaphthoquinone, In *Molecules,* v.22, 1631-1643.

[26] Ferreira, P. G.; Cruz, R. S.; Costa, V. G.; Ferreira, V. F.; Futuro, D. O. 2017. Development and Characterization of Emulgels for

Treatment of Chronic Wounds, In *Asian Journal of Pharmaceutical Technology Innovation,* v.5, 40-48.

[27] Campos, V. R.; Gomes, A. T. P. C.; Cunha, A. C.; Neves, M. D. G. P. M. S. Ferreira, V. F; Cavaleiro, J. A. S. 2017. Efficient access to alfa vinylporphyrin derivatives via palladium cross coupling of alpha-bromoporphyrins with tosylhydrazones, In *Beilstein Journal of Organic Chemistry,* v.13, 195-202.

[28] Cardoso, M. F. C., Gomes, A. T., Moreira, C. D., Simões, M. M., Neves, M. G., Rocha, D. R., Da Silva, F. C., Moreirinha, C., Almeida, A., Ferreira, V. F., Cavaleiro, J. A. S.; 2017. Efficient Catalytic Oxidation of 3-Arylthio- and 3-Cyclohexylthio-lapachone Derivatives to New Sulfonyl Derivatives and Evaluation of Their Antibacterial Activities, In *Molecules,* v.22, E302.

[29] Oliveira, L. F. G.; Souza-Silva, F.; Cysne-Finkelstein, L.; Rabelo, K.; Amorim, J. F.; Azevedo, A. S.; Bourguignon, S. C.; Ferreira, V. F.; Paes, M. V.; Alves, C. 2017. Evidence for Tissue Toxicity in BALB/c Exposed to a Long-Term Treatment with Oxiranes Compared to Meglumine Antimoniate, In *Biomed Research International,* v.2017, 1-11.

[30] Santos, T. F.; De Jesus, J. B.; Neufeld, P. M.; Jordão, A. K.; Campos, V. R.; Cunha, A. C.; Castro, H. C.; De Souza, M. C. B. V.; Ferreira, V. F.; Rodrigues, C. R.; Abreu, P. A. 2017. Exploring 1,2,3-triazole derivatives by using in vitro and in silico assays to target new antifungal agents and treat Candidiasis, In *Medicinal Chemistry Research,* v.26, 680-689.

[31] Ferreira, P. G.; Caroline, N. D.; Da Silva, F. C.; Ferreira, V. F. 2017. Green Synthetic Routes to Pharmaceutical Drugs, In *Current Green Chemistry,* v.3, 259-276.

[32] Sodero, A. C. R.; Abrahim-Vieira, B.; Torres, P. H. M.; Pascutti, P. G.; Garcia, C. R. S.; Ferreira, V. F.; Rocha, D. R.; Ferreira, S. B.; Silva Jr, F. P. 2017. Insights into cytochrome bc1 complex binding mode of antimalarial 2-hydroxy-1,4-naphthoquinones through molecular modelling, In *Memórias do Instituto Oswaldo Cruz,* v.112, 299-308.

[33] Da Silva, F. C.; Ferreira, V. F. 2017. March for Science in Brazil, In *Revista Virtual de Química,* v.9, 474-475.

[34] Pereira, A. M. V. M.; Lacerda, P. S. S.; Silva, B. N. M.; Neves, M. G. P. M. S.; Silva, A. M. S.; Silva, B. V.; Da Silva, F. C.; Ferreira, V. F.; Pinto, A. C.; Cavaleiro, J. A. S. 2017. One-pot synthesis of new isatin-porphyrin conjugates by the palladium Buchwald-Hartwig methodology involving alpha-amino porphyrinatonickel(II) and 3-ketal isatin derivatives, In *Dyes and Pigments,* v.139, 247-254.

[35] Nicoletti, C. D.; Ferreira, V. F.; Cruz, R. S.; Futuro, D. O. 2017. Papain Hydrogel with Polysorbate: Critical Micelle Concentration and Thermodynamics Stability, In *Asian Journal of Pharmaceutical Technology & Innovation,* v.5, 1-9.

[36] Teixeira, R. I.; Dos'santos, I. C.; Garden, S. J.; Carneiro, P. F.; Ferreira, V. F.; De'lucas, N. C. 2017. Photosensitizing Properties of 6H -Dibenzo[b,h]xanthene Derivatives, In *Chemistry Select.*, v.2, 11732-11738.

[37] Moreira, C. S.; Silva, A. C. J. A.; Novais, J. S.; Sá Figueiredo, A. M.; Ferreira, V. F.; Da Rocha, D. R.; Castro, H. C. 2017. Searching for a potential antibacterial lead structure against bacterial biofilms among new naphthoquinone compounds, In *Journal of Applied Microbiology*, v.122, 651-662.

[38] Novais, J. S.; Campos, V. R.; Silva, A. C. J. A.; De Souza, M. C. B. V.; Ferreira, V. F.; Keller, V. G. L.; Ferreira, M. O.; Dias, F. R. F.; Vitorino, M. I.; Sathler, P. C.; Santana, M. V.; Resende, J. A. L. C.; Castro, H. C.; Cunha, A. C. 2017. Synthesis and antimicrobial evaluation of promising 7-arylamino-5,8-dioxo-5,8-dihydroiso-quinoline-4-carboxylates and their halogenated amino compounds for treating Gram-negative bacterial infections, In *RSC Advances,* v.7, 18311-18320.

[39] Cardoso, M. F. C; Forezi, L. S. M.; Cavalcante, V. G. S.; Juliani, C. S. R.; Resende, J. A. L. C.; Rocha, D. R.; Da Silva, F. C.; Ferreira, V.F. 2017. Synthesis of New Xanthenes Based on Lawsone and Coumarin via a Tandem Three-Component Reaction, In *Journal of the Brazilian Chemical Society,* v.28, 1926-1936.

[40] Da Silva, T. B., Da Silva, F. C.; Ferreira, V. F. 2017. Synthetics Methods for the Preparation of Biaryls, In *Revista Virtual De Química,* v.9, 1258-1284.

[41] Montes, G. C., Da Silva, B. N. M., Rezende, B., Sudo, R. T., Ferreira, V. F., De Da Silva, F. C, Pinto, Ac., Da Silva, B. V., Zapata-Sudo, G. 2017. The Hypnotic, Anxiolytic, and Antinociceptive Profile of a Novel μ-Opioid Agonist, In *Molecules,* v.22, 800.

[42] Ferreira, P. G.; Da Silva, F. C.; F. Ferreira, V. F. 2017. The Importance of Chemistry for the Circular Economy, In *Revista Virtual de Química,* v.9, 452-473.

[43] Martinez T., S.; Ferreira, V. F. 2017. The Isatins of Professor Angelo, In *Revista Virtual de Química*, v.9, 1154-1183.

[44] Gonzaga, D. T. G., Ferreira, L. B. G., Moreira, M. C. T. E., Von Ranke, N. L., Anastácio, P. F. P., Sposito, S. A. P., Arruda, J. C., Dantas, L. P., De Freitas, H. R., De Melo Reis, R. A., Penido, C., Bello, M. L., Castro, H. C., Rodrigues, C. R., Ferreira, V. F., Faria, R. X., Da Silva, F. C. 2017. 1-Aryl-1H- and 2-aryl-2 H-1,2,3-triazole derivatives blockade P2X7 receptor in vitro and inflammatory response in vivo, In *European Journal of Medicinal Chemistry*, 139, 698-717.

[45] Forezi, L. S. M.; Cardoso, M. F. C.; Costa, D. C. S.; Da Silva, F. C.; Ferreira, V. F. 2017. 2,3-Dichloro-1,4-Naphthoquinone in Organic Synthesis: Recent Advances, In *Mini-Reviews in Organic Chemistry,* v.14, 375-390.

[46] Gonzaga, D. T. G.; Da Silva, F. C.; Ferreira, V. F.; Wardell, J. L.; Wardell, S. M. S. V. 2016. Crystal Structures of 1-Aryl-1H- and 2-Aryl-2H-1,2,3-triazolyl Hydrazones. Conformational Consequences of Different Classical Hydrogen Bonds, In *Journal of the Brazilian Chemical Society,* v.27, 2322-2333.

[47] Gonzaga, D. T. G.; Da Silva, F. C.; Ferreira, V. F.; Wardell, J. L.; Wardell, S. M. S. V. 2016.Crystal Structures of 2-Phenyl-2H-1,2,3-Triazol-4-Carbaldehyde, an Active beta-Glycosidase Inhibition Agent, and (1-Phenyl-1H-1,2,3-Triazol-4-yl)Methyl Benzoate and (2-

(4-Fluorophenyl)-2H-1,2,3-Triazole-4-yl)Methanol, Two Moderately Active Compounds, In *Journal of Chemical Crystallography,* v.46, 67-76.

[48] Carneiro, P. F.; Pinto, M. C. F. R.; Marra, R. K. F.; Campos, V. R.; Resende, J. A. L. C.; Delarmelina, M.; Carneiro, J. W. M.; Lima, E. S.; Da Silva, F. C.; Ferreira, V. F. 2016. Insight into and Computational Studies of the Selective Synthesis of 6H-Dibenzo[b,h]xanthenes, In *Journal of Organic Chemistry,* v.81, 5525-5537.

[49] De Souza, I. C. A.; Faro, L. V.; Pinheiro, C. B.; Gonzaga, D. T. G.; Da Silva, F. C.; Ferreira, V. F.; Miranda, F. S.; Scarpellini, M.; Lanznaster, M. 2016. Investigation of cobalt(III)-triazole systems as prototypes for hypoxia-activated drug delivery, In *Dalton Transactions,* v.45, 13671-13674.

[50] Da Silva, F. C.; Ferreira, V. F. 2016. Natural Naphthoquinones with Great Importance in Medicinal Chemistry, In *Current Organic Synthesis,* v.13, 334-371.

[51] Ferreira, V. F.; Nicoletti, C. D.; Ferreira, P. G.; Futuro, D. O.; Da Silva, F. C. 2016. Strategies for Increasing the Solubility and Bioavailability of Anticancer Compounds: alfa-Lapachone and Other Naphthoquinones, In *Current Pharmaceutical Design,* v.22, 5899-5914.

[52] Carneiro, P. F.; Pinto, M. C. R. F.; Marra, R. K. F.; Da Silva, F. C.; Resende, J. A. L. C.; Rocha E Silva, L. F.; Alves, H. G.; Barbosa, G. S.; De Vasconcellos, M. C.; Lima, E. S.; Pohlit, A. M.; Ferreira, V. F. 2016. Synthesis and antimalarial activity of quinones and structurally-related oxirane derivatives, In *European Journal of Medicinal Chemistry,* v.108, 134-140.

[53] Moura L. A, De Almeida A. C., Da Silva A. V., De Souza V. R., Ferreira V. F., Menezes M. V., Kaiser C. R., Ferreira S. B., Fuly A. L. 2016. Synthesis, Anticlotting and Antiplatelet Effects of 1,2,3-Triazoles Derivatives, In *Medicinal Chemistry,* v.12, 733-741.

[54] Louvis, A. R.; Silva, N. A. A.; Semaan, F. S.; Da Silva, F. C.; Saramago, G.; De Souza, L. C. S. V.; Ferreira, B. L. A.; Castro, H.

C.; Salles, J. P.; Souza, A. L. A.; Faria, R. X.; Ferreira, V. F.; Martins, D. L. 2016. Synthesis, characterization and biological activities of 3-aryl-1,4-naphthoquinones - green palladium-catalysed Suzuki cross coupling, In *New Journal of Chemistry,* v.40, 7643-7656.

[55] Cardoso, M. F. C.; Forezi, L. S. M.; Da Silva, F. C.; Pinto, A. C.; Neves, M. G. P. M. S.; Ferreira, V. F.; Cavaleiro, J. A. S. 2016. Synthetic methodologies leading to porphyrin-quinone conjugates, In *Journal of Porphyrins and Phthalocyanines, v.*20, 167-189.

[56] Silva, B. N. M.; Pinto, A. C.; Silva, F. C.; Ferreira, V. F.; Silva, B. V. 2016. Ultrasound-Assisted Synthesis of Isatin-Type 5'-(4-Alkyl/Aryl-1H-1,2,3-triazoles) via 1,3-Dipolar Cycloaddition Reactions, In *Journal of the Brazilian Chemical Society*, v.27, 2378-2382.

[57] Delarmelina, M.; Ferreira, G. B.; Ferreira, V. F.; Carneiro, J. W. M. 2016. Vibrational spectroscopy of lapachol, beta- and alfa-lapachone: Theoretical and experimental elucidation of the Raman and infrared spectra, In *Vibrational Spectroscopy,* v.86, 311-323.

[58] Cunha, A. C.; Jordão, A. K.; Souza, M. C. B. V.; Ferreira, V. F.; Almeida, M. C. B.; Wardell, J. L.; Tiekink, E. R. T. 2016. 1-Anilino-5-methyl-1H-1,2,3-triazole-4-carbaldehyde, In *IUCrData,* Volume 1, Part 1, January,| x160038.

In: Heterocycles
Editor: Jonas M. Danielsen

ISBN: 978-1-53617-804-3

Chapter 2

RECENT ADVANCES IN THE SYNTHESIS OF FIVE- AND SIX-MEMBERED OXYGEN HETEROCYCLES AND THEIR APPLICATION IN NATURAL PRODUCT SYNTHESIS

Ngangbam Renubala Devi and Anil K. Saikia*
Department of Chemistry, Indian Institute of Technology Guwahati, Guwahati, Assam, India

ABSTRACT

Heterocycles are the basic core structure in a plethora of naturally occurring bioactive molecules. Oxygen containing heterocyclic compounds such as dihydro-, tetrahydro-, pyrano-pyrans and tetrahydrofurans are the key building blocks of various natural and biologically active products. Methodologies such as Prins cyclization, oxonium-ene, hetero-Diels-Alder, metal catalyzed cyclization, 1,n-enyne rearrangement reactions are commonly practiced for the synthesis of above heterocycles. To mention a few, decytospolide A, which has tetrahydropyran as its core structure can be synthesized using Prins

* Corresponding Author's E-mail: asaikia@iitg.ac.in

cyclization reaction and has antibacterial and antitumour properties. (+)-Muconin has both tetrahydro-pyran and –furan rings and shows potent *in vitro* cytotoxicity against PACA-2 (pancreatic cancer) and MCF-7 (breast cancer). Saikia and co-workers synthesized bioactive substituted tetrahydropyrans *via* oxonium-ene reaction and natural product (+)-civet *via* Prins cyclization reaction. In 2009, Krause and co-workers reportedly synthesized natural products (+)-linalool oxide, (-)-isocyclocapitelline and (-)-isochrysotricine using gold catalyst.

As such, significant methodologies are reported for the synthesis of dihydropyrans and tetrahydropyrans separately over the years, but reports solely concentrated on oxygen heterocycles as a whole and their reaction methodologies and applications are not reported yet. This review will highlight recent developments (since 2014) in reaction strategies for the synthesis of dihydro-, tetrahydro-, pyrano-pyrans and tetrahydrofurans and their applications in natural product synthesis.

Keywords: heterocycles, carbocycles, dihydropyran, tetrahydropyran, pyranopyrans, tetrahydrofuran, benzopyran, isochromane, natural product, Prins cyclization, oxonium-ene, hetero-Diels-Alder, enyne rearrangement, metal catalyzed cyclization, macrolide, cytotoxicity, Alzeimer's disease, Parkinsonism, homoallyl alcohol, aldehyde, retrosynthesis, stereoselectivity, regioselectivity and diastereoselectivity.

1. Introduction

1.1. Heterocyclic Compounds

Heterocyclic compounds constitute one of the most important classes of organic compounds and they are widely distributed in nature. They are vital for the existence of living beings. A cyclic organic compound containing all carbon atoms in ring formation is referred to as a *carbocyclic compound.* If at least one atom other than carbon forms a part of the ring system, then it is designated as a *heterocyclic* compound. Nitrogen, oxygen and sulfur are the most common heteroatoms but heterocyclic rings containing other hetero atoms are also widely known. An enormous number of heterocyclic compounds are known and this number is increasing rapidly. Heterocyclic compounds may be classified into *aliphatic* and *aromatic*. The aliphatic heterocycles are the cyclic analogues of amines, ethers, thio-ethers, amides, etc. These compounds generally consist of small (3- and 4-membered) and medium size (5 to 7 membered) ring systems. The aromatic heterocyclic compounds, in contrast, are those which have a heteroatom in the ring and behave in a manner similar to benzene in some of their properties. Furthermore, these compounds also comply with the general rule proposed by Hückel. Besides the vast distribution of heterocycles in natural products, they are also the major components of biological molecules such as DNA and RNA. More than half of all known organic compounds are heterocycles. They are present in a wide variety of drugs, most vitamins, many natural products, biomolecules, and biologically active compounds, including antitumor, antibiotic, anti-inflammatory, antidepressant, antimalarial, anti-HIV, antimicrobial, antibacterial, antifungal, antiviral, antidiabetic, herbicidal, fungicidal, and insecticidal agents. Also, they have been frequently found as a key structural unit in synthetic pharmaceuticals and agrochemicals. Some of these compounds exhibit a significant solvatochromic, photochromic, and biochemi-luminescence properties. Most of the heterocycles possess important applications in materials science such as dye stuff, fluorescent sensor, brightening agents, information storage, plastics, and analytical reagents. In addition, they have applications in supra molecular and polymer chemistry, especially in conjugated polymers. Moreover, they act as organic conductors, semiconductors, molecular wires, photovoltaic cells, and organic light-emitting diodes (OLEDs), light harvesting systems, optical data carriers, chemically controllable switches, and liquid crystalline compounds. Heterocycles are also of considerable interest because of their synthetic utility as synthetic intermediates, protecting groups, chiral auxiliaries, organo catalysts and metal ligands in asymmetric synthesis. Therefore, a substantial attention has been paid to develop efficient new methods to synthesize heterocycles.

Heterocycles are able to get involved in an extraordinarily wide range of reaction types. Some are readily oxidized, but resist reduction, while others can be readily hydrogenated but are stable toward the action of oxidizing agents. The ability of many heterocycles to produce stable complexes with metal ions has great biochemical significance. Five and six membered oxygen heterocyclic compounds are important in synthetic chemistry because of their immense presence in various natural products and pharmaceuticals. They are the core units of various bio-active molecules. It is well explored that pyran and furan derivatives showed numerous types of pharmacological and biological properties, such as fungicidal, insecticidal, acaricidal activity, antiviral, anticonvulsant and antileishmanial activity. Moreover, they have been introduced as non-peptide human immunodeficiency virus (HIV) protease inhibitors. For example, Centrolobine is a tetrahydropyran containing biomolecule, isolated from the heart-wood of the tree *Centrolobiumtomentosum*, has antibiotic and anti-fungal activities. Whereas its enantiomer (+)-centrolobine occurs in the closely related species *Centrolobiumrobustum* also found to have similar properties [1]. Martiriol which has both dihydropyran and pyranopyran rings in its core unit, is isolated from the red algae *Laurencia* and shows potent activity against various tumor cells [2]. (+)-Muconin is a tetrahydropyran-tetrahydrofuran core and α,β-unsaturated γ-lactone containing metabolite, which has potent cytotoxicity against pancreatic and breast cancer cells. It was first isolated by McLaughlin and co-workers from the leaves of *Rollinia mucosa* [3]. These oxygen containing heterocycles such as dihydro-, tetrahydro-, pyrano-pyrans and tetrahydrofurans can be synthesized by using various strategies, Prins cyclization, oxonium-ene, hetero-Diels-Alder, metal catalyzed cyclization, 1,n-enyne rearrangement reactions. The above mentioned heterocycles are briefly discussed in this chapter along with their synthetic applications leading to the total synthesis of some of the bioactive molecules.

1.2. Dihydropyrans

Various oxygen containing heterocycles are abundantly found in nature. Most of the pharmacological activities are attributed to the substitution pattern on the heterocyclic rings. Five and six membered oxygen containing heterocycles possess significant biological properties among them dihydropyrans are widely spread in natural products or biologically active molecules. For example, dihydropyrans such as Laulimalide (**1**) is a cancer therapeutic lead recently isolated from pacific marine sponges [4]. Aspergillide C (**2**) is a 14-membered macrolide which has a property of cytotoxity, firstly derived from the marine fungus *Aspergillus ostianus* also contains dihydropyran ring [5]. Xenicin (**3**) is a diterpenoid having a nine membered carbocyclic ring. It is a marine natural product obtained from the soft coral *Xenia elongate* and has significant cytotoxic and antibacterial properties [6]. Similarly, diterpenes kalihipyrans A (**4**) and B (**5**) derived from marine sponge *Acanthella carvernosa*, are also used in antifouling activity [7] (*Figure 1.2.1*).

1.3. Pyranopyrans

Pyranopyrans are the class of bicyclic compounds containing oxygen on either side of the ring. They also found to possess many medicinal and pharmacological properties. The glycofused benzopyran (**6**) was recently identified as a novel ligand for amyloid β peptidase in the development of novel therapeutics for Alzheimer's disease [8]. Martiriol (**7**) which has both dihydropyran and pyranopyran rings is isolated from the red algae of the genus Laurencia and shows potent activity against various tumor cells [2]. Similarly, Compound (**8**) is also a natural product produced by a rare bacterial strain *Actinoalloteichus nanshanensis sp. Nov* [9]. Salinomycin (**9**) is an antibiotic that has recently been found to activate cell death in a variety of human cancer cells. It is a polyether made by a strain of *streptomyces albus*. Streptomyces is a type of bacteria, usually found in soil but sometimes in plants and animals as well [10]. Calyxin J and Epicalyxin J (**10**) are also pyranopyran containing heterocyclic biomolecules which show antitumor activities [11] (*Figure 1.3.1*).

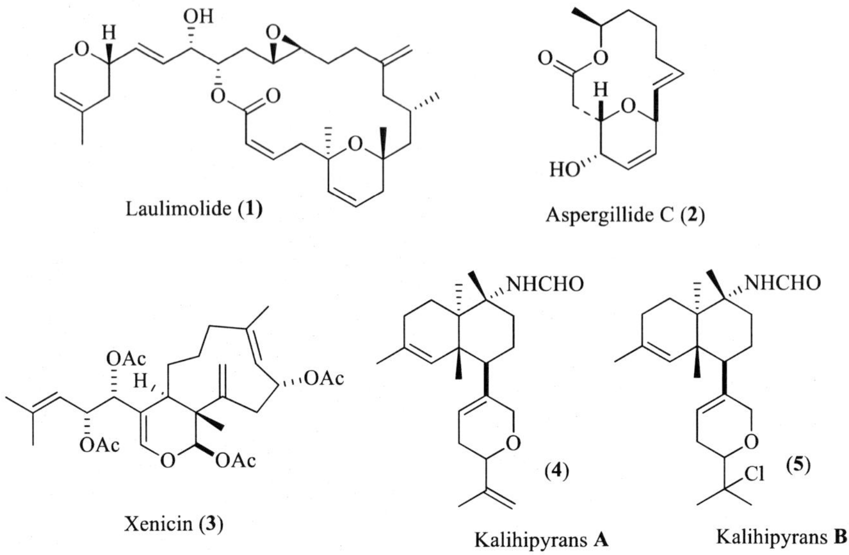

Figure 1.2.1. Bioactive molecules containing dihydropyrans.

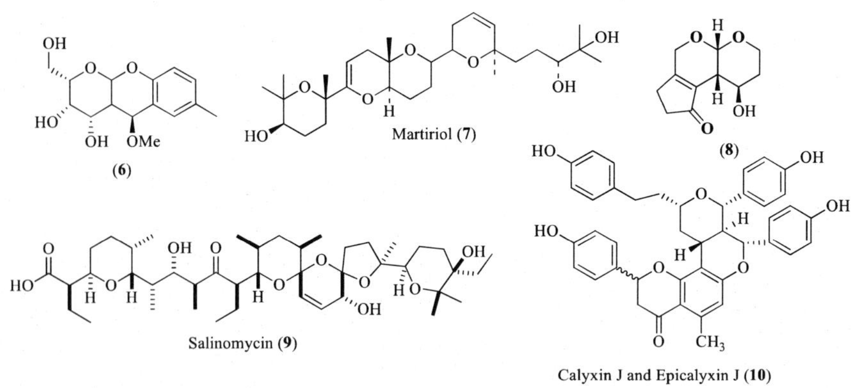

Figure 1.3.1. Biologically active compounds containing pyranopyran rings.

1.4. Tetrahydropyrns

Tetrahydropyran, a six membered oxygen containing saturated class of compounds, are found as core structural units in many naturally occurring bioactive mlecules. For example, tetrahydropyran containing natural product, pyranicin (**11**) acts as antimalarial, antiparasitic and antitumor drug and it has recently exhibited promising results against Parkinson's disease [12]. Similarly, another tetrahydropyran containing natural product calyxin L (**12**) isolated from *Alpinia blepharocalyx* seeds, is used for the treatment of stomach disorders in Chinese medicine and also acts as potent member of the calyxin class and showed activity against human HT-1080 fibrosarcoma and murine 26-L5 carcioma [13]. Bryostatin 1 (**13**) is a representative marine macrolide natural product that contains tetrahydropyran rings as substructures and is isolated by the Pettit and coworkers from the marine bryozoan *Bugula neritina* and structurally identified by an X-ray crystallographic analysis [14]. Bryostatin 1 is a potent and isoform-selective protein kinase C activator and shows anti-cancer, anti-Alzheimer's disease, and anti-HIV activities (*Figure 1.4.1*). Because of these significant biological activities and synthetically challenging complex molecular architecture, a number of renowned synthetic chemists devoted their efforts to the development of concise synthesis of bryostatin 1 and its congeners.

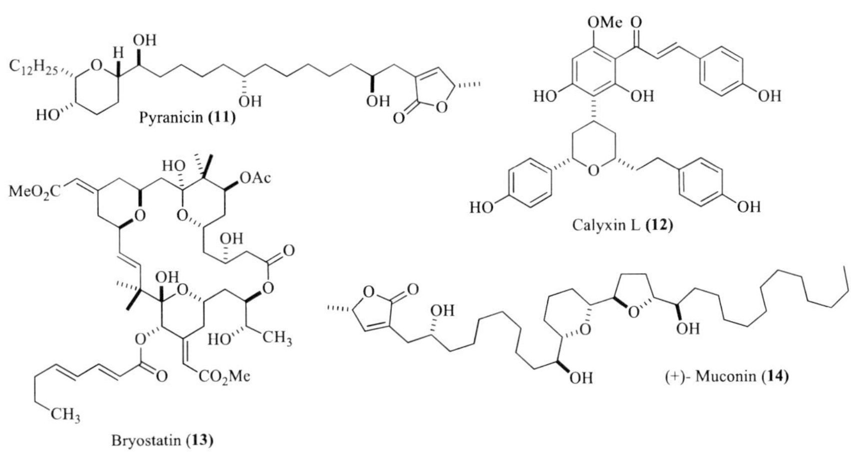

Figure 1.4.1. Biologically active compounds containing tetrahydropyran rings.

1.5. Tetrahydrofurans (THF)

Tetrahydrofurans are oxygen containing five-membered saturated compounds. They are widely found in nature and also show significant biological properties. Some tetrahydrofuran core containing biomolecules are shown in *Figure 1.5.1*. For example, a 16 membered macrodiolide pamamycin-607 (**15**), isolated from *streptomyces albonigerand S. aurantiacus* shows potent activity against gram-positive bacteria as well as acts against phytopathogenic fungi [15]. The mycotoxin citreoviridine (**16**), isolated from *Penicillum citreoviride*, is a potent inhibitor of the soluble mitochondrial ATPase and also has tetrahydrofuran ring as core unit [16]. Aflatoxin (**17**) another tetrahydrofuran containing moety is found to be highly potent carcinogenic metabolites and produced by fungi *Aspergillusflavus* and *Aspergillus parasiticus* [17]. Muscarine (**18**), a furan containing biologically active alkaloid is the major toxic principle found in the well-known mushroom *Amanita muscaria* (Fly agaric) [18]. It acts as a selective agonist of the neurotransmitter acetylcholine on smooth muscles of the gastrointestinal tract, eye exocrine glands and heart. Stolonoxide are a group of compounds containing *trans*-2,5-disubstituted-tetrahydrofuran in its core unit. Stolonoxide A and B (**19**) have cyclic peroxy and unsaturated 15 carbon side chain as its substitutes along with THF ring. The major difference between A and B, being the *cis-trans* olefinic positions in the unsaturated side chain [19]. The family of stolonoxide has high potent anti-cancer properties. (+)-Chamuvarinin (**20**) is a potent anticancer and antitrypanosomal agent, which has tetrahydropyran ring linked to two tetrahydrofuran rings (THF-THF), isolated from the root of *Uvaria chamae* by Laurens and co-workers [20]. Nonactin (**21**), an ionophoric macrolide antibiotic isolated from a variety of Streptomyces cultures has been shown to possess antitumor activity both against mammalian cell lines *in vitro* and against Crocker sarcoma 180 in studies with mice [21] (*Figure 1.5.1*).

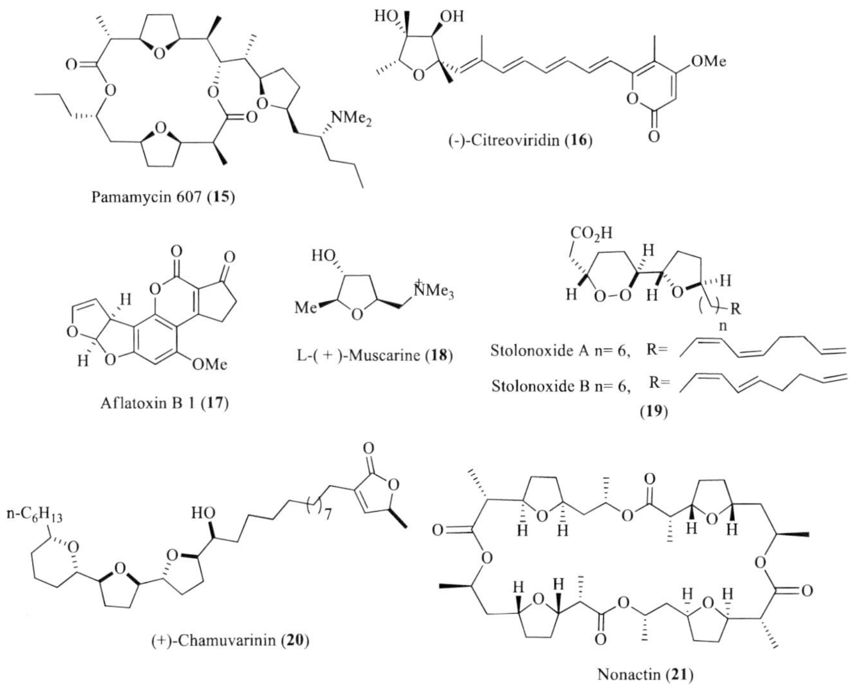

Figure 1.5.1. Biologically active compounds containing tetrahydrofurans.

1.6. Benzopyrans/Isochromanes

Another important class of cyclic ethers is benzopyrans/isochromanes in which a pyran ring is fused with a benzene ring. The isochromane fragment is the core of various natural compounds, which exhibit a wide variety of pharmacological activities which are promising for the treatment of migraines, Parkinson's disease and schizophrenia [22]. Penidicitrinim B (**22**) isolated from *Penicillium citrinum strains*, containing a C-1 phenyl substituted isochromane ring possesses antioxidant properties [23]. Similarly, a wide spread 1-Aryl-6,7-dimethoxyisochromanes shows a wide range of biological activities such as analgesic, muscle relaxant, anti-depressant, anti-inflammatory, anti-histaminic, anti-coagulant, hypotensive with peripheral and central activities, and are adrenergic antagonists [24]. Biologically active compound **23** containing an isochromane ring shows affinity towards Ca-receptors [26]. PNU-109291 (**24**) an isochromane ring containing moety is selective 5-HT1D receptor agonist with anti-migraine potential, like sumatriptan [25] (*Figure 1.6.1*).

Similarly, pongapinone (**25**) and deguelin (**26**) have benzopyran as their core unit. While **25** acts as inhibitor of interleukin-1 production [26] and **26** as inhibitors of electron transport acting at NADH-ubiquinone oxidoreductase [27] (*Figure 1.6.1*).

2. Synthetic Methodologies

2.1. Prins Cyclization Reaction

The Prins reaction is one of the fundamental methods for C-C bond formation, first reported by H. J. Prins in the year 1919. In his initial studies, H. J. Prins performed the reaction of formaldehyde (**1**) and styrene (**2**) in aqueous acidic medium, resulting in a mixture of products such as diol **3**, unsaturated alcohol **4** and 1,3-dioxane (**5**) (*Scheme 1*), thus becoming one of the prominient synthetic tools for the generation of heterocyclic compounds [28].

In 1955, Hanschke further developed Prins reaction for the selective synthesis of tetrahydropyran (THP) rings **8** by combining 3-buten-1-ol (**6**) with a variety of aldehydes or ketones **7** in the presence of acid, called Prins cyclization reaction (*Scheme 2*) [29].

In most of the cases the Prins cyclization reaction is highly diastereoselective and give 2,4,6-substituted tetrahydropyrans with all equatorial substitutions. The diastereoselectivity can be explained by considering the mechanism of the reaction [30]. The lone pair of the homoallylic alcohol **10** attacks the carbonyl centre of the aldehyde **9** in the presence of acid to generate an oxocarbenium ion **12** as a key intermediate, which undergoes 6-endo-trig-cyclization to give more stable tetrahydropyranyl cation **13**. According to Alder's DFT calculations, carbocation **13** in its chair conformation is stabilized by stereoelectronic effects. The C2- C3 and C5-C6 σ^* and σ orbital overlap both the equatorial lone pair of the oxygen atom and the vacant p orbital at C4. Optimal overlap is reached when the hydrogen atom at C4 is pseudo-axial. This stabilization favors equatorial attack by the nucleophile to give tetrahydropyran **14** (*Scheme 3*).

Subba Reddy and coworkers reported a stereoselective synthesis of tetrahydropyran and tetrahydrofuran core **18** of the biomolecule (+)-muconin *via* Prins cyclization reaction. The starting materials homoallyl alcohol **15** and aldehyde **16** were treated with TFA in dichloromethane to give hydroxylated THP-THF containing moety **17** as given in *Scheme 4* [31].

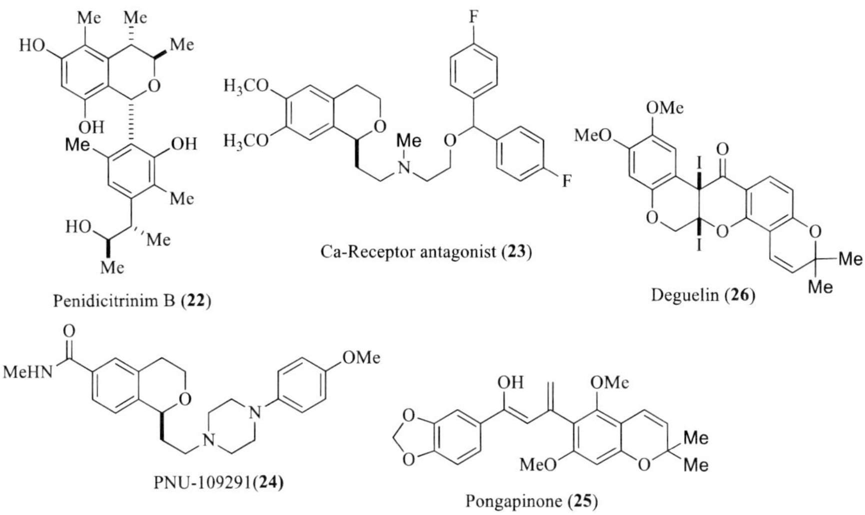

Figure 1.6.1. Biologically active compounds containing benzopyrans/isochromanes.

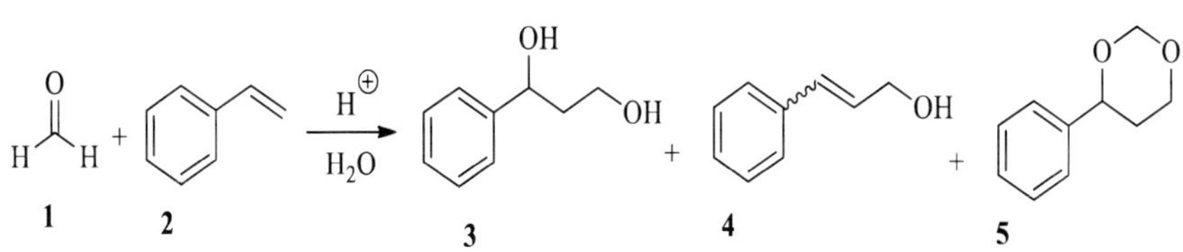

Scheme 1. Prins reaction.

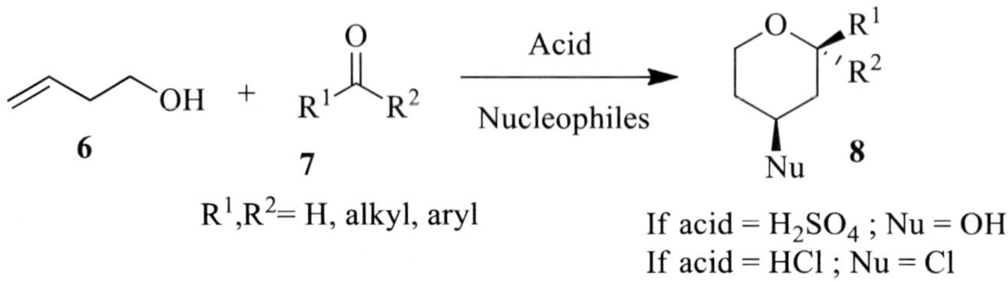

Scheme 2. Prins cyclization reaction.

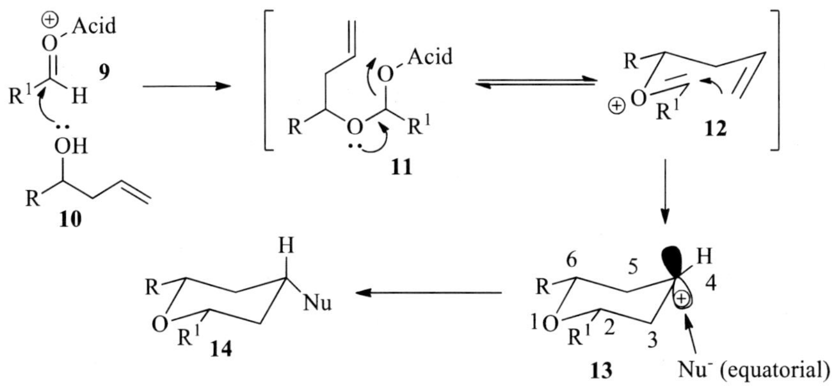

Scheme 3. Mechanism of Prins cyclization reaction.

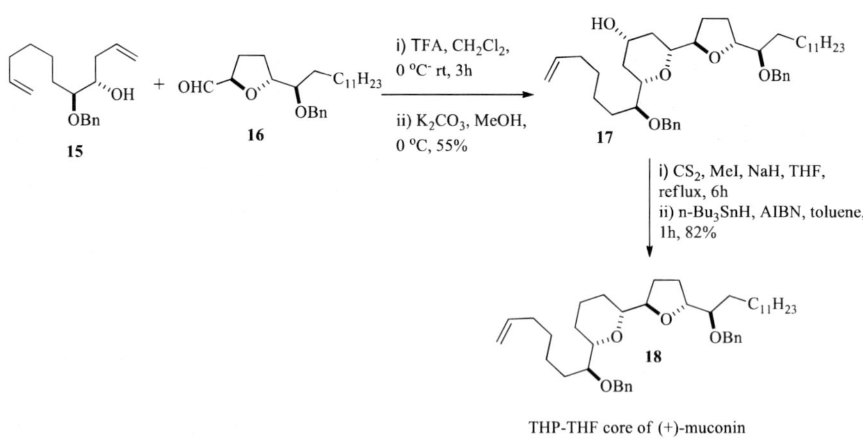

Scheme 4. Synthesis of THP-THF core of (+)- muconin.

Dehydroxylation of compound **17** led to the formation of basic core structure of (+)-muconin. It can be mentioned that the homoallylic alcohol **15** can be easily synthesized from the commercially available hept-6-ene-1-ol using a sequence of reactions such as Wittig olefination, Sharpless epoxidation, etc.

The retrosynthetic analysis of (+)-muconin is shown below in *scheme 5*. From this point of view (+)-muconin **22** can be easily synthesized by coupling of the above fragment **18** along with butenolide fragment **21** *via* cross-metathesis.

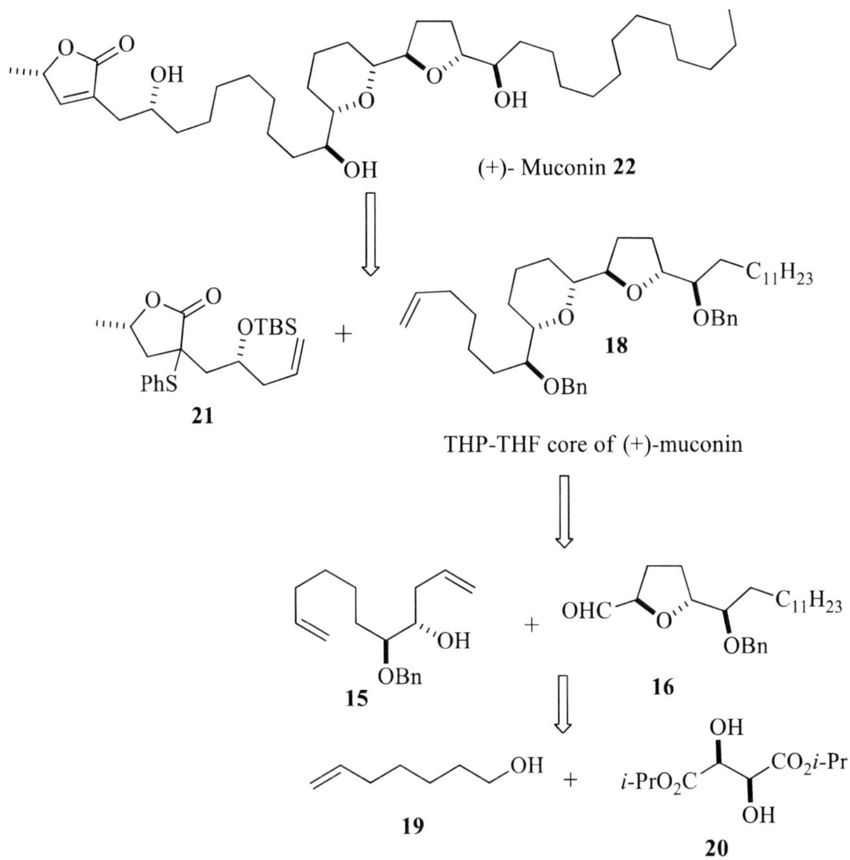

Scheme 5. Retrosynthetic aspects of (+)- muconin.

In 2018, Wang and coworkers reported a 7 step total synthesis of cochlearoid B (**26**) by utilizing Suzuki-Mayura cross coupling and acid mediated Prins cyclization reactions (*Scheme 6*) [32]. They first started with the ring closure of secondary chloride **23** in the presence of Zn dust, a Barbier type reaction condition, to produce a diastereomeric mixture of bridged bicyclic alcohols **24a** and **24b**. After the methodology was established, the precursor biaryl **25** was then treated with (+)-CSA in such a way that in a single final step, two MOM protecting groups were removed as well as dehydration between the carbonyl and the proximal phenol occurred to form an intermediate oxocarbenium B which was then trapped by the double bond to give cochlearoid B as a single isomer. Cochlearoids are a class of meroterpenes, isolated from *Ganoderma cochlear* and found to have significant pharmacological properties of neurological disorder [33].

Vankar and coworkers also recently reported a Lewis acid catalyzed silyl-Prins/alkyne-Prins cyclization reactions for the stereoselective synthesis of 1,2- annulated sugars having substituted tetrahydropyran **29** or tetrahydrofuran **32** scaffolds by using substituted alkenyl **27** or alkynyl alcohols **30** respectively along with aldehydes **28** and **31** (*Scheme 7*). These annulated molecules are almost identical to those of marine toxins such as maitotoxin which could act as useful building block for the synthesis of this type of natural product [34].

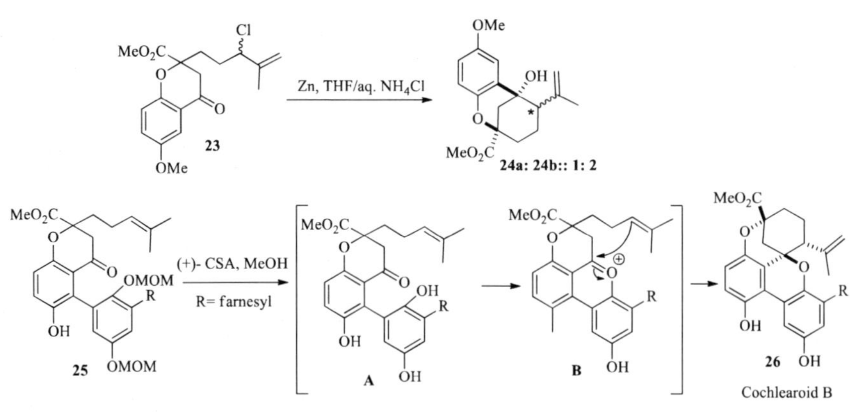

Scheme 6. Synthesis of cochlearoid B.

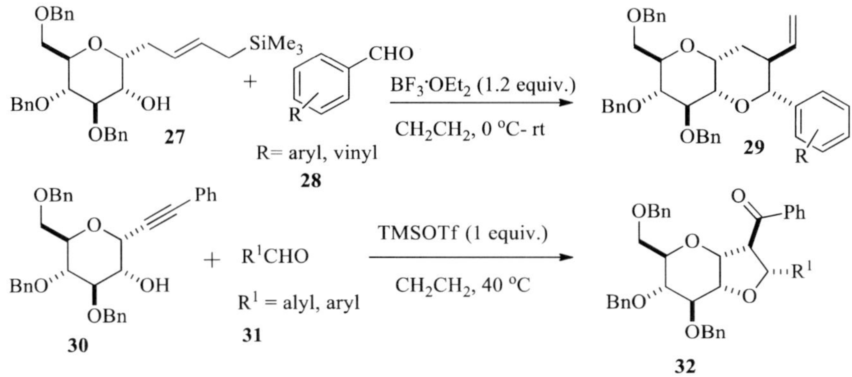

Scheme 7. Silyl-Prins/alkyne-Prins cyclization reaction.

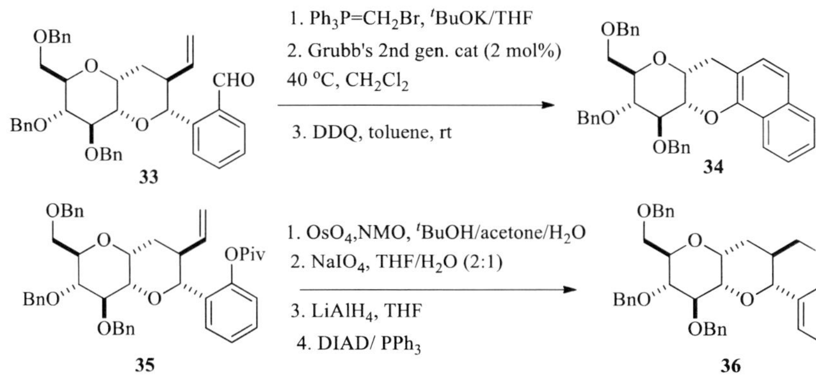

Scheme 8. Synthesis of cis-sugar fused 4H-naptho [1,2-b] pyran and cis-sugar fused pyrano[3,2-c] [1] benzopyran.

As a synthetic utility of this methodology the substrates **33** and **35** were further examined and led to the synthesis of *cis*-sugar fused 4*H*-naptho[1,2-*b*] pyran **34** and *cis*-sugar fused pyrano[3,2-c][1]benzopyran **36** derivatives, respectively (*Scheme 8*). Wittig olefination of compound **33** followed by ring closing metathesis with Grubbs 2nd generation catalyst and oxidation by DDQ afforded cis-sugar fused 4*H*-naptho [1,2-*b*] pyran compound **34**. On the other hand, 1,2-annulated sugar **35** was subjected to dihydroxylation by OsO_4 followed by cleavage of the diol with $NaIO_4$ to give an aldehydic intermediate which after reduction with $LiAlH_4$ and intramolecular cyclization produced *cis*-sugar fused pyrano [3,2-*c*][1]benzopyran **36**.

Kantee and coworkers reported a concise synthesis of tetrahydropyranyl diarylheptanoids **39** and **40** which have tetrahydropyran (THP) rings by using aldehydes **37** and homoallyl alcohols **38** in the presence of per-rhenic acid in dichloromethane at room temperature as shown in *Scheme 9*. The most important aspect is the utilization of Prins cyclization reaction to construct the tetrahydropyran cores [35]. Under diarylheptanoids, natural products such as Centrolobine **41** and Rhoiptelol B **42** (*Figure 1.1.1*) can be cited as important examples. These classes of alkaloids exhibit a wide range of biological properties such as anti-inflammatory, antioxidant, anticancer, antibacterial and antiosteoporotic activities [36].

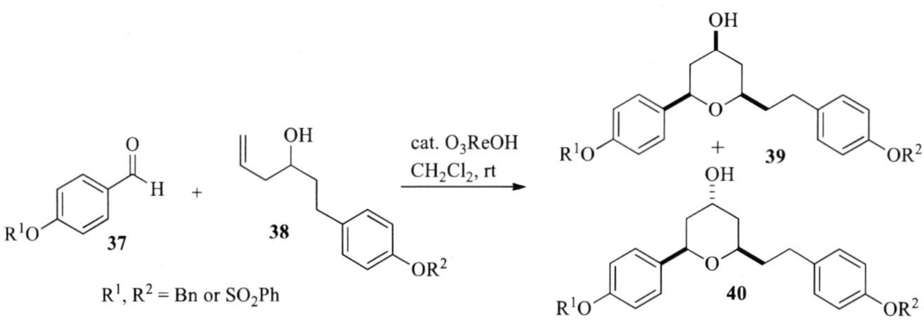

Scheme 9. Synthesis of tetrahydropyranyl diarylheptanoids.

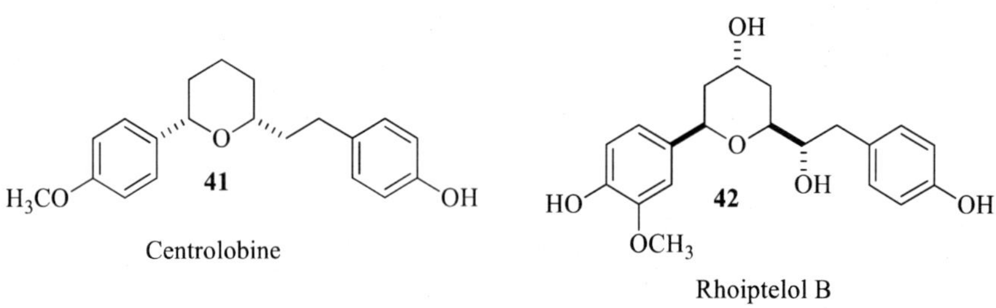

Figure 1.1.1. Structure of Centrolobine and Rhoiptelol B.

The same methodology has been utilized for the syntheis of natural products **46-47** and **50-51**. The synthesis was procceded by using chiral homoallylic alcohols **43** and **48** with aldehyde **44** *(Scheme 10)*. Previously, these diarylheptanoids **46**, **47**, **50** and **51**, containing THP cores were isolated from *Dioscorea villosa* and *Dioscorea nipponica* [37].

A Prins bicyclization methodology for the stereoselective synthesis of *trans*-fused hexahydropyrano [3,4-*c*] chromene derivatives **54** has been developed from bis-homoallyl alcohols 52 and aldehydes **53** by Subbha Reddy and coworkers (*Scheme 11*). The synthetic versatility of this approach has been extended for the synthesis of calyxin I and J analogues. Calyxin family of terpenoids exhibit antiproliferative activity toward HT-1080 fibrosarcoma and colon 26-L5 carcinoma and they have a pyranochromene ring as one of main structural motif [38].

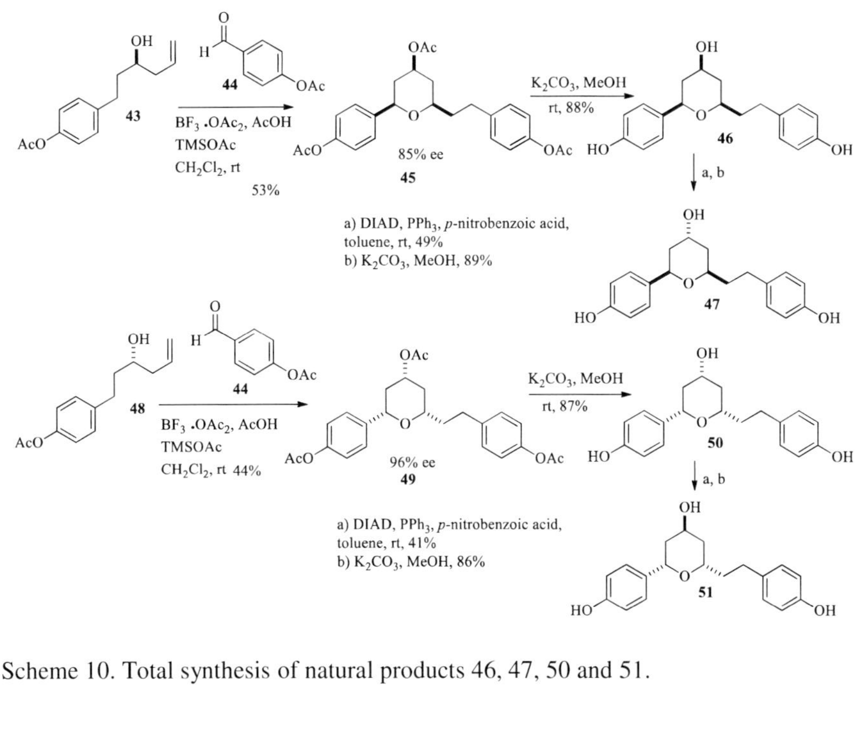

Scheme 10. Total synthesis of natural products 46, 47, 50 and 51.

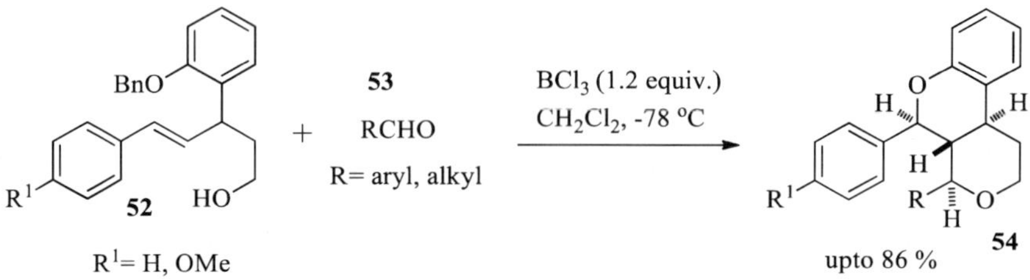

Scheme 11. Synthesis of trans-fused hexahydropyrano [3,4-c] chromene derivatives.

This methodology was successfully applied for the syntheses of pyranochromene ring of calyxin I (**58**), (*Figure 1.1.2*) and J (**10**), (*Figure 1.4.1*) (*Scheme 12*). The starting material, γ,δ-unsaturated alcohol **55** was treated with benzaldehyde **56** in the presence of 1.2 equiv. BCl_3 at – 78°C to produce compound **57**, the core structure of calyxins I and J, in 65% yield.

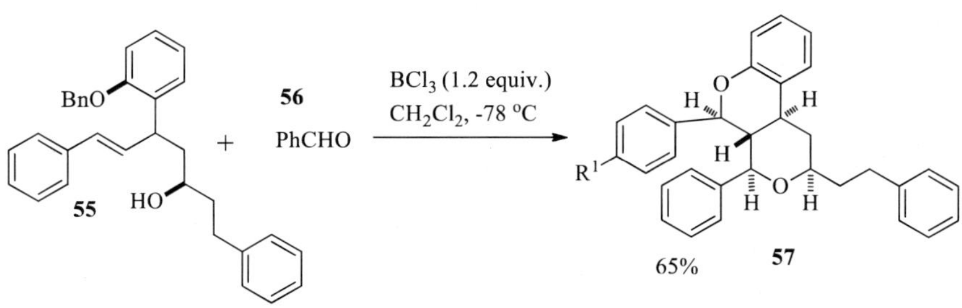

Scheme 12. Syntheses of pyranochromene.

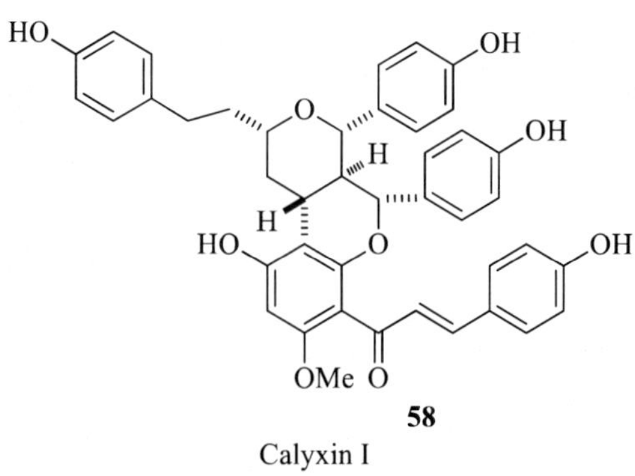

Figure 1.1.2. Structure of calyxin I.

In 2017, Ma and coworkers utilized *β*-allenols for the construction of pyran rings *via* Prins cyclization reaction. It was reported that, in the presence of $In(OTf)_3$ 5,5-disubstituted 3,4-allenols **59** and aldehydes **60** underwent cyclization reaction to synthesize 3,6-dihydro-2*H*- pyrans **61** in dichloromethane at 45°C (*Scheme 13*) [39].

A new and diastereoselective synthesis of activated 2,3,4,6-tetrasubstituted tetrahydro-2*H*-pyrans **64** was described by Padrón and coworkers. In this reaction, *bis*-homoallylic alcohols **62** and aldehydes **63** were used as substrates for S_N2'–Prins type cyclization tandem process leading to the formation of tetrahydopyran derivatives, catalyzed by iron(III) chloride, with three new stereocenters in a single step as shown in *Scheme 14* [40].

Saikia and coworkers reported a triflic acid mediated Prins cyclization of homopropargylic alcohols **65** with aldehydes **66** to give 3,6-dihydro-2*H*-pyran-4-yl trifluoromethanesulfonates **67** in high yields (*scheme 15*). The dihydropyran thus formed can be post functionalized into different 4-alkyl and aryl substituted products forming **68**, **69**, **70** and **71** *via* Suzuki, Sonogashira, Stille and Heck coupling reactions, respectively [41].

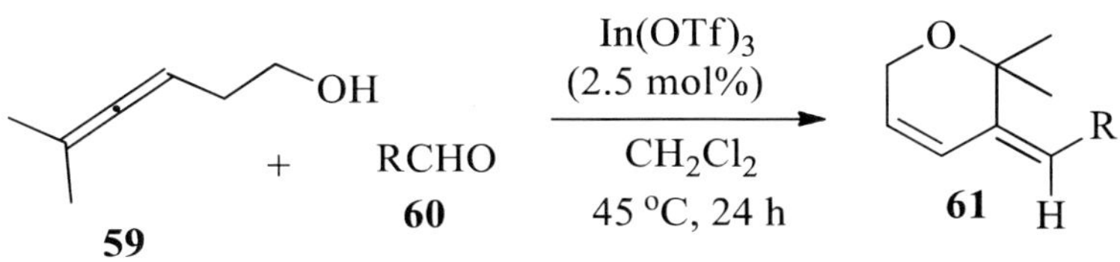

Scheme 13. Synthesis of dihydropyrans.

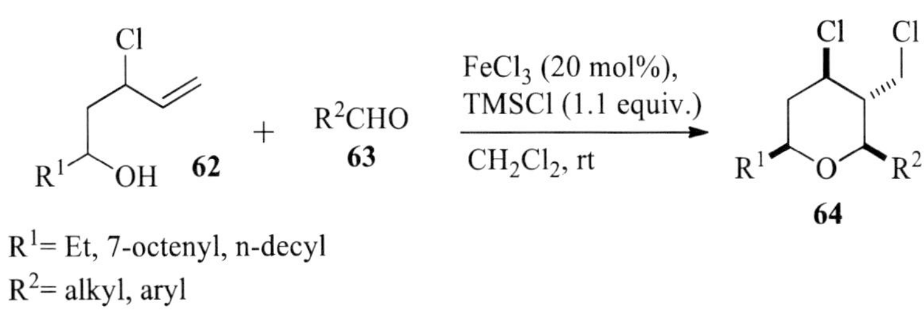

Scheme 14. Syntheiss of tetrahydropyrans.

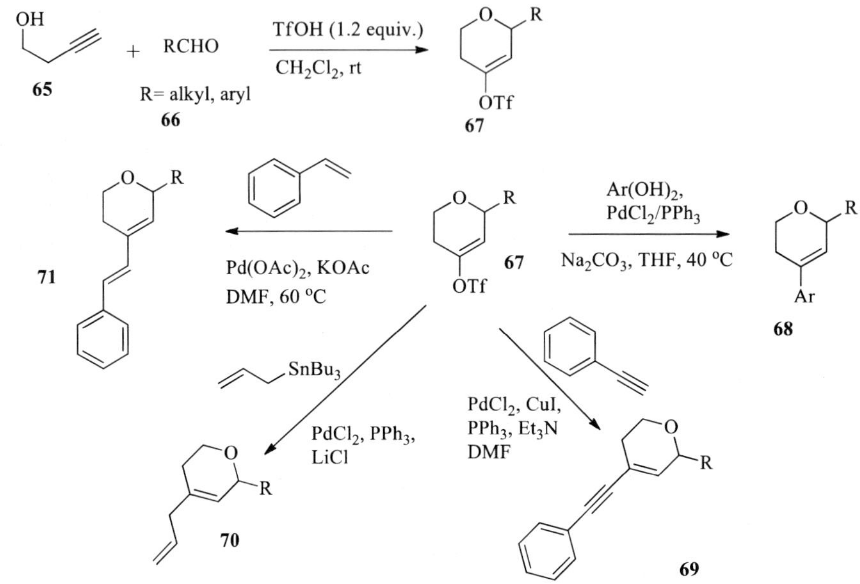

Scheme 15. Synthesis of substituted dihydropyrans.

A new class of highly acidic confined iminoimidodiphosphate (iIDP) Brønsted acids **L**1 was developed by List and coworkers and later utilized in catalytic asymmetric Prins cyclization of both aliphatic and aromatic aldehydes **73** with homoallyl alcohols **72** as described in *scheme 16.* Olefin functionalized 4- methylenetetrahydropyrans **74** were obtained in good to excellent yields and with good to excellent regio- and enantioselectivities [42].

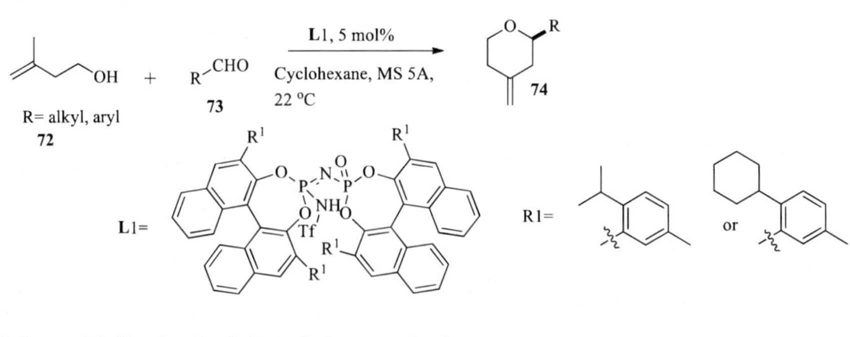

Scheme 16. Synthesis of 4-methylenetetrahydropyrans.

In 2019, Saikia and coworkers reportedly synthesized hexahydrofuro[3,4-*b*]furan-4-ol **78** and its dimer **77** with by using ((4*S*,5*R*)-2,2-dimethyl-5-vinyl-1,3-dioxolan-4-yl)methanol **75** and aldehydes **76** as substrates (*Scheme 17*). A variety of substituted furanol and its bis-derivatives were obtained in good yields with excellent diastereo- and enantio-selectivity. The reaction was mediated by borontrifluoride etherate whereby the substrates underwent a tandem Prins and Pinacol rearrangement. The dimer can be further converted into its corresponding monomer using aqueous zinc(II) chloride in THF in quantitative yields [43].

In 2014, Saikia and coworkers established an indium triflate catalyzed Prins cyclization reaction of acrylyl enol ethers **79** to give tetrahydrofuran ring **80** stereo-and regioselectively in good yields [44] (*Scheme 18*).

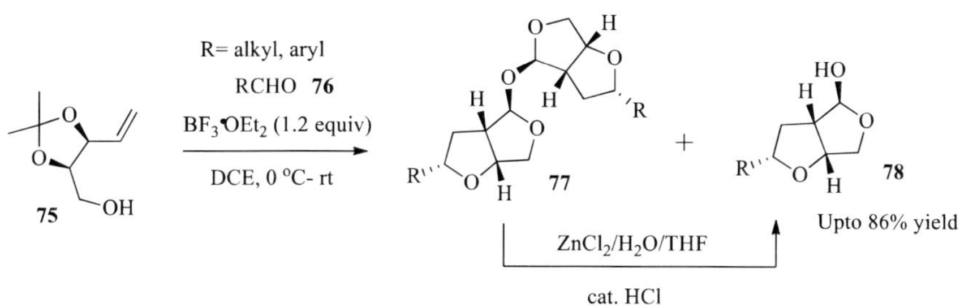

Scheme 17. Synthesis of hexahydrofuro[3,4-b]furan-4-ol.

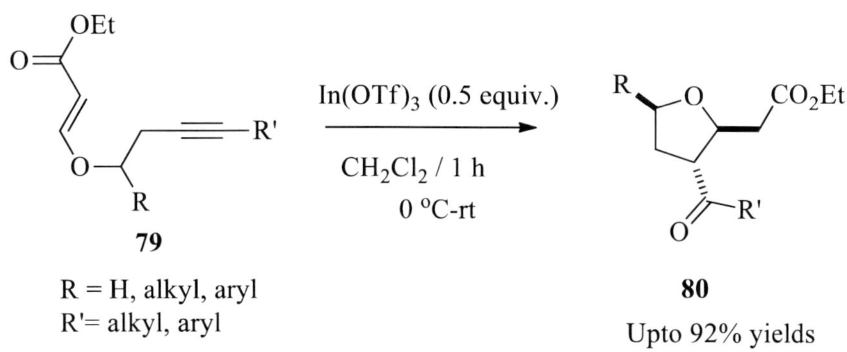

Scheme 18. Synthesis of tetrahydrofurans.

Saikia and coworkers also recently utilized triethylsilyl homoallylic alcohols **81** and aldehydes **82** for the synthesis of dihydropyrans **83** in highly diastereo- and regio-selective manner *via* silyl Prins cyclization reaction. The reaction was catalyzed by borontrifluoride etherate ($BF_3 \cdot OEt_2$) [45] (*Scheme 19*).

Enol ethers **84** were again exploited for the intramolecular Prins cyclization in the presence of *p*-toluenesufonic acid to give substituted 4-*o*-tosyl-tetrahydropyrans 85 in good yields by the same group (*Scheme 20*) [46]. The obtained substituted tetrahydropyrans can also be functionalized to give corresponding 4- iodotetrahydropyrans **86**, para-substituted tetrahydropyrans with azide **87** and detosylation to give reduced tetrahydropyrans **88**.

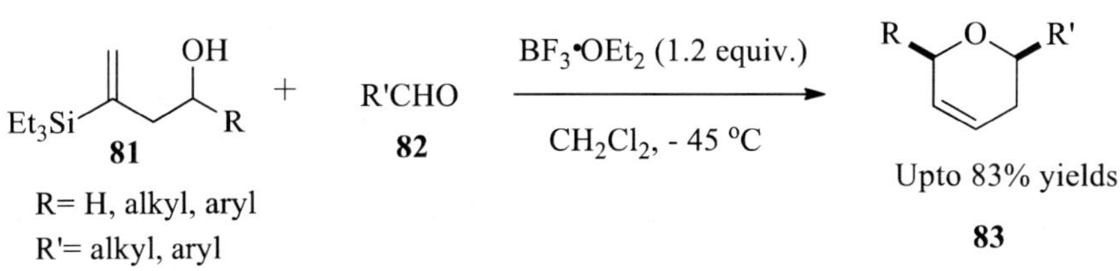

Scheme 19. Synthesis of dihydropyrans.

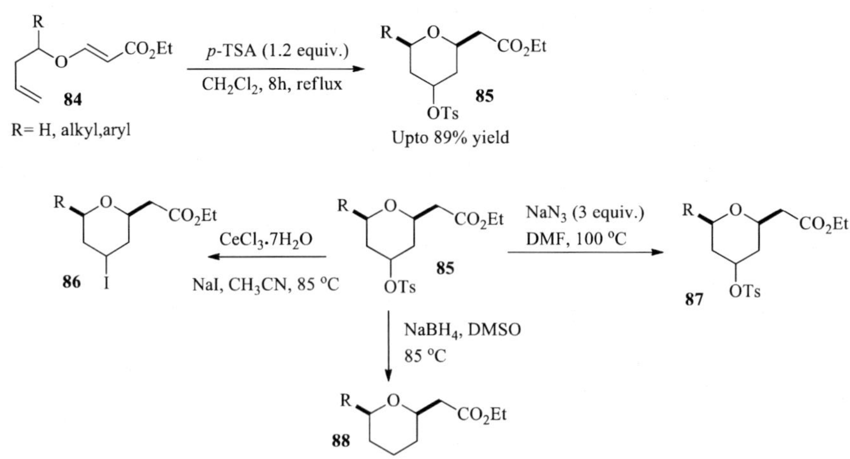

Scheme 20. Prins cyclization of enol ether leading to tetrahydropyrans.

2.2. N- Enyne Rearrangement Reaction

Enyne is an organic compound which has both double and triple bonds together in conjugation or separated by one or more carbon or any other heteroatom. Depending on the mechanistic pathways of the transition-metal-catalyzed cycloisomerization reaction of enynes, different pathways can be considered based on the reaction conditions and on the choice of the precatalyst. Generally, it involves formation of metal complex to either alkene or alkyne or both which allows the activation of either both moieties or only one of them. Three main mechanisms can be proposed (*Scheme 21*) [47]. In pathway a, simultaneous complexation of both alkene and alkyne can occur, leading to the formation of metallacycle I as an intermediate. However, presence of a functional group in the allylic position can also allow the formation of a π-allyl complex II (pathway b) which could further react with the triple bond. Finally, in pathway c hydrometalation of the alkyne led to the corresponding vinylmetal III which is reactive enough to allow the carbometalation of the olefin.

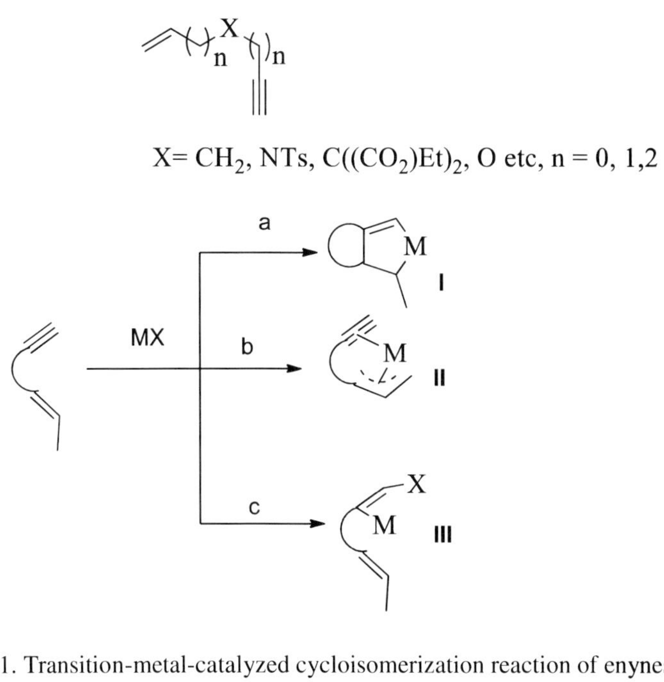

Scheme 21. Transition-metal-catalyzed cycloisomerization reaction of enynes.

These transition-metal-catalyzed cyclization of 1,n-enynes has become an atom-economical and powerful method for building diverse carbo- and hetero-cyclic frameworks. This strategy includes metal-catalyzed oxidative cyclization of 1,n-enynes, which offers several advantages over the traditional approaches as numerous new functional groups can be easily introduced into the ring systems and functionalization in the alkene and alkyne moieties occur in both positions of the π bond. Some examples involving metal catalyzed cyclization reaction for the synthesis of five and six oxygen containing heterocycles are described below.

Two novel 16-oxa-vitamin D_3 analogues **95a** and **95b** were reportedly synthesized by Okamoto and coworkers, using a series of reactions, such as Ti(II)-mediated enyne cyclization, Cu-catalyzed allylation, Ru-catalyzed ring-closing metathesis, and a low-valent titanium (LVT)-mediated stereoselective radical reduction of epoxide [48]. The 16-oxa-bicyclic-ring precursor **92** was prepared through a Ti(II)-mediated cyclization reaction of enyne **89**, followed by an *in situ* Cu-catalyzed allylation by allyl bromide to give triene **91**. A ring-closing metathesis of the triene yielded in the formation of oxa-bicyclic compound **92**. Finally coupling of compound **93** with **94a/b** produced the desired compound 16-oxa-1α,25-$(OH)_2VD_3$ **95a** and 16-oxa-19-*nor*-1α,25-$(OH)_2VD_3$ **95b** respectively, in the presence of palladium catalyst. (*Scheme 22*)

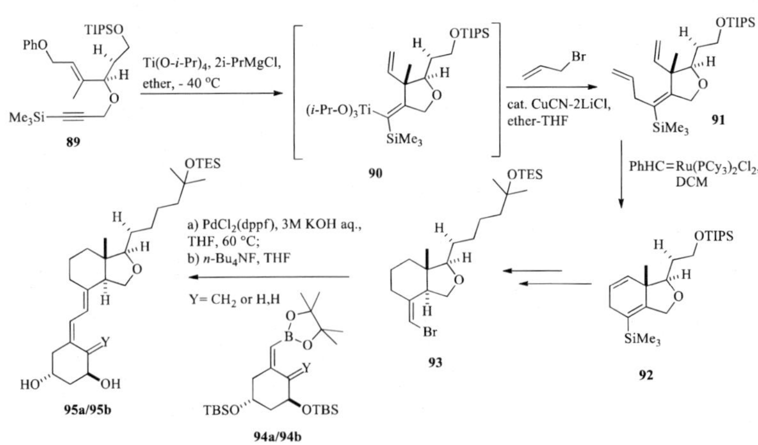

Scheme 22. Synthesis of 16-oxa-vitamin D3 analogues.

Eupalinilide E **99** promotes the expansion of human HSPCs (hematopoietic stem and progenitor cells) and inhibits subsequent differentiation, leading to increased numbers of clinically useful cells. In 2016, Siegel and coworkers reported an enantioselective synthesis of eupalinilide E **99** starting from (*R*)-carvone (**96**) as shown in *scheme 23* [49]. Key transformations enabling the approach include Dess-Martin oxidation, a diastereoselective borylative enyne cyclization, a late-stage double allylic C−H oxidation, Luche reduction and aluminum-mediated diastereoselective epoxidation reactions.

Yang and co-workers synthesized tetrahydrofuran derivatives **102** by reductive cyclization of *O*-tethered 1,6-enynes **100** in moderate to good yields. A precatalyst of $FeCl_2$ and iminopyridine **101** which was activated *in situ* by a combination of diethylzinc and magnesium bromide etherate was used for the cyclization reaction as shown in *Scheme 24* [50].

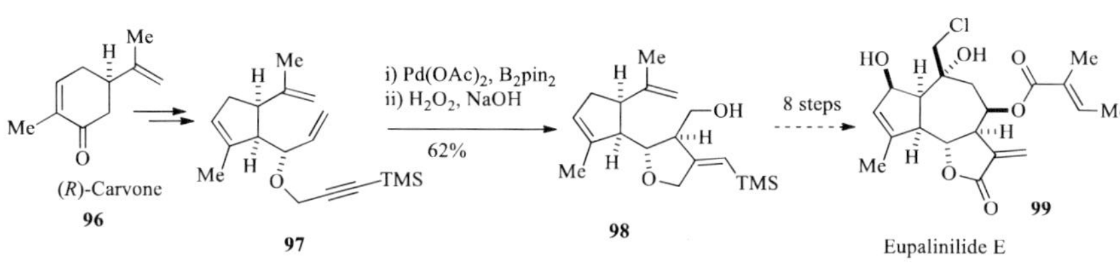

Scheme 23. Synthesis of eupalinilide E.

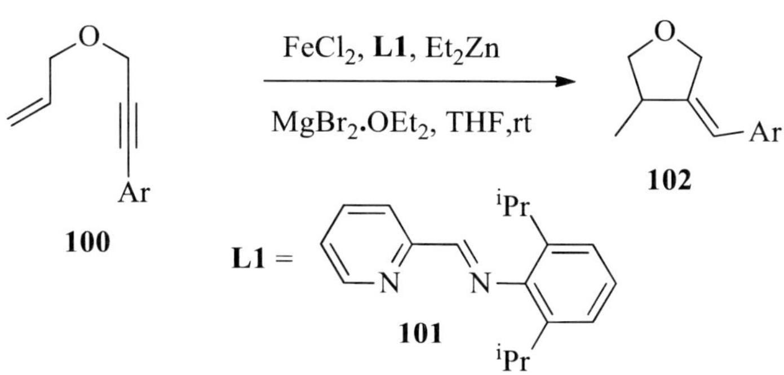

Scheme 24. Synthesis of tetrahydrofurans.

2.3. Hetero-Diels-Alder Reaction

Hetero Diels-Alder (HDA) reaction is a perfect mechanism for the synthesis of heterocyclic compounds. Many of them used Lewis/Bronsted acid in order to obtain polyheterocyclic compounds. However, many of these synthetic methods suffer from drawbacks in reaction conditions, yields, regio- and diastereo-selectivity. So, reactions that can overcome these situations are still preferred and some methodologies that previously reported are discussed below.

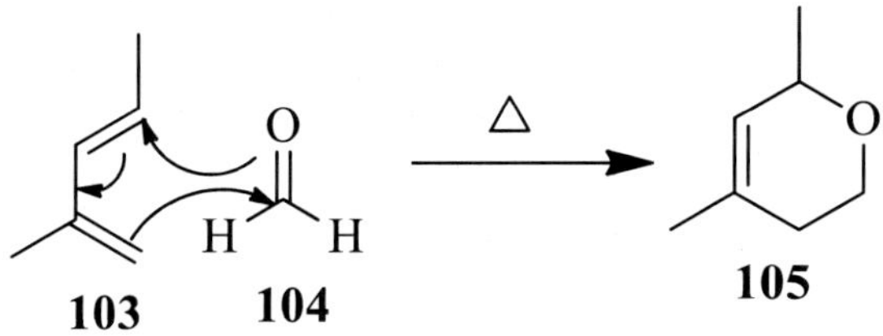

Scheme 25. Diels-Alder reaction of diene and aldehyde.

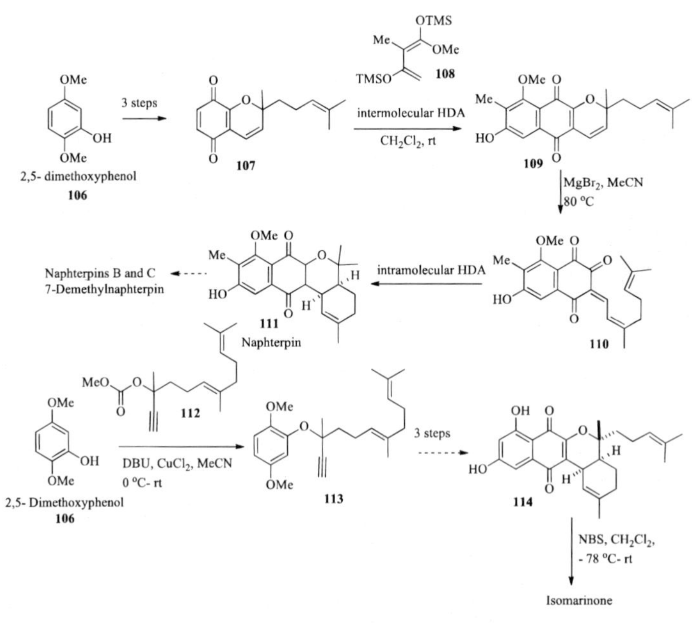

Scheme 26. Synthesis of 7-demethylnaphterpin, naphterpin B, C and isomarinone.

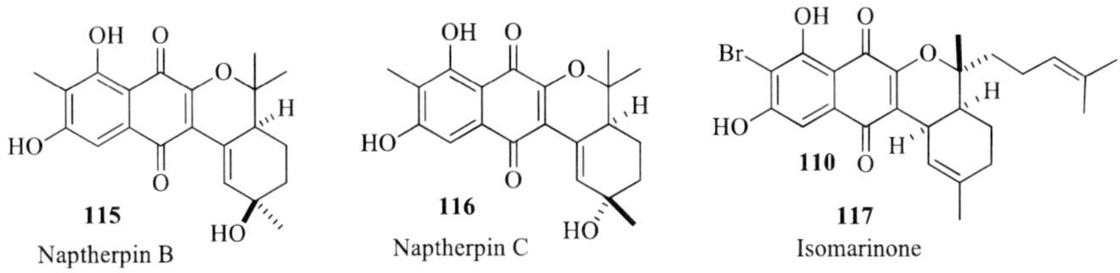

Figure 1.3.1. Structures of naphterpin B, naphterpin C and isomarinone.

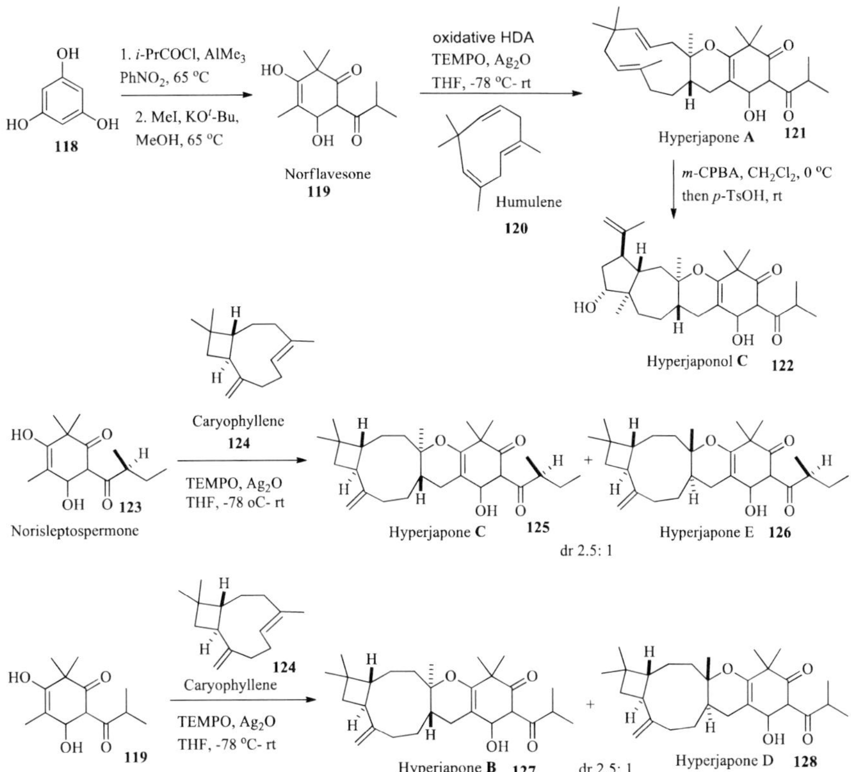

Scheme 27. Synthesis of hyperjapones A,B,C,D and E and hyperjaponol C.

HDA reaction was first reported by Gresham and Steadman in 1951. In their report they have prepared 2,4-dimethyl-5,6-dihydro-1,2-pyran **105** by reacting methylpentadiene **103** with formaldehyde (**104**) in the presence of heat [51] (*Scheme 25*). The reaction was assumed to be unusual, as in this case, formaldehyde was acting as a dienophile.

Recently, a concise and step-wise strategy for the total synthesis of naphthoquinone meroterpenoids such as the naphterpins and marinone has been developed by George and coworkers (*Scheme 26*). The approach features a sigmatropic rearrangement, an electrocyclization, intermolecular as well as intramolecular hetero-Diels-Alder reactions. All the syntheses were proceeded with 2,5-dimethoxyphenol (**106**) as the starting material. The strategy resulted in the divergent synthesis of natural products such as 7-demethylnaphterpin, naphterpin B, naphterpin C and isomarinone (*Figure 1.3.1*) all synthesized for the first time [52].

Hyperjapones and hyperjaponols are complex natural biomolecules isolated from *Hypericum japonicum*. In 2016, George and coworkers reported a biosynthetic route for the synthesis of hyperjaones A-E and hyperjaponol C *via* oxidative hetero Diels–Alder reaction using norflavesone **119**, norisoleptospermone **123**, humulene **120** and caryophyllene **124** as starting materials under different reaction conditions as depicted in *scheme 27* [53].

Psiguajdianone, a caryophyllene-derived meroterpenoid dimer, was isolated from *Psidium guajava*. It is used in traditional medicines; the leaves of this plant have long been used for the treatment of diabetes, cough, spasms, diarrhea, and hyperglycemia, etc. [54]. In 2019, Xei and coworkers were able to isolate Psiguajdianone **139**, a novel caryophyllene-derived meroterpenoid dimer, isolated from *Psidium guajava* and performed a total synthesis of the same by using cascade Knoevenagel condensation/hetero Diels−Alder reaction and dimerization as shown in *Scheme 28* [55]. The intermediate **132** was synthesized by three-component one-pot reaction with phloroglucinol derivative **129**, formaldehyde **130** and caryophyllene **124** in the presence of silica−$HClO_4$ (50% w/w), but resulted to give two major compounds **131** and **133**, which were also isolated from the leaves of *P. guajava* and named as guapsidial A (**131**) and psiguajadial D (**133**) respectively along with two minior compounds **132** and **134**. The mixture of **131**, **132**, **133** and **134** could not be separated and purified, further treatment of the mixture of products with $NaBH_3CN$ reduced the corresponding the aldehydes to give **135**, **136**, **137** and **138** respectively. Dimerization of compound **136** by single electron oxidant AgOAc in CH_3CN gave the desired natural product **139** in 57% yields.

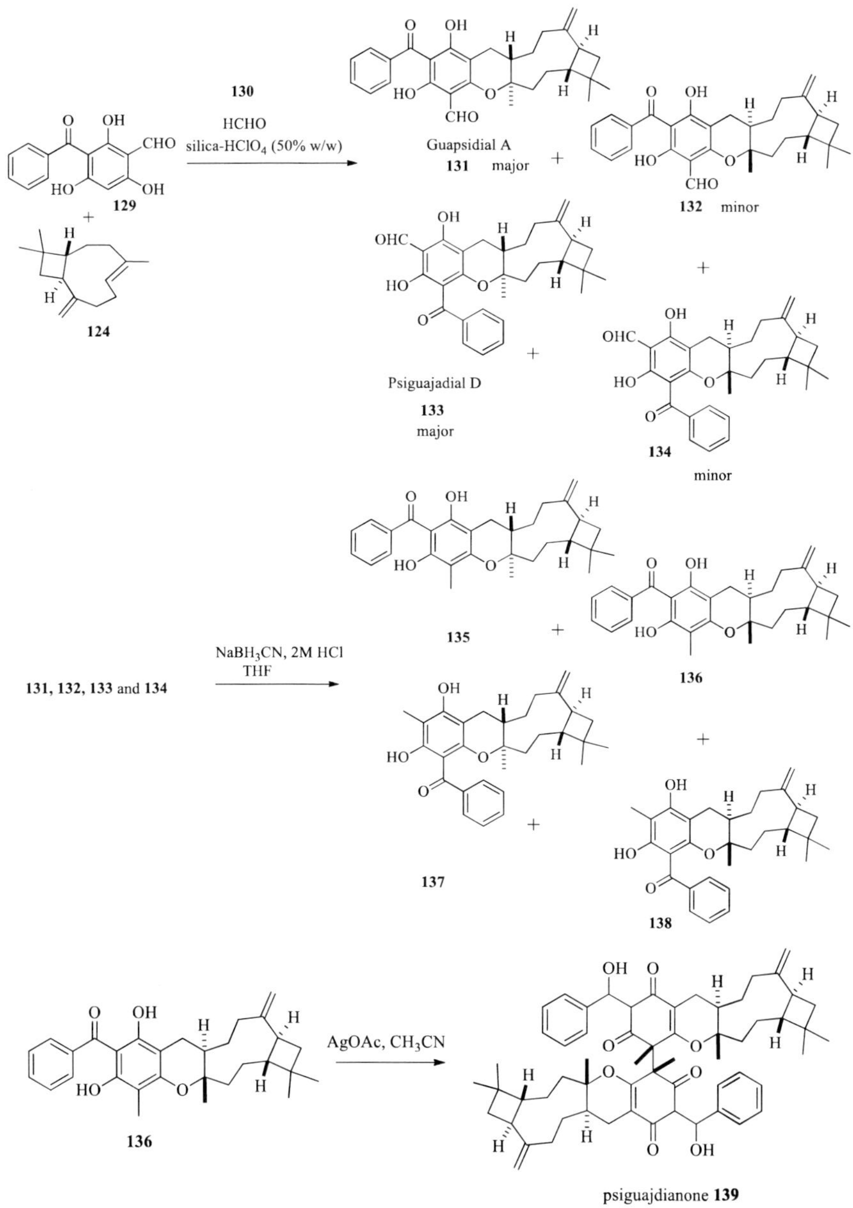

Scheme 28. Synthesis of Psiguajdianone.

Inulavosin **144** was isolated from the roots of *Inula nervosa* and showed a wide range of biological activities such as antibacterial, piscicidal, antifungal and melanogenesis inhibitory properties [56]. Wua and coworkers reportedly synthesized inulavosin (**144**) from commercially available and inexpensive *m*-cresol **140**. Hydrogenolysis of **141** followed by reaction with methyllithium produced salicyl alcohol derivative **143** in 83% overall yield, which underwent hetero Diels–Alder dimerization in the presence of gallium(III) triflate to give **144** in 88% yield [57] (*Scheme 29*).

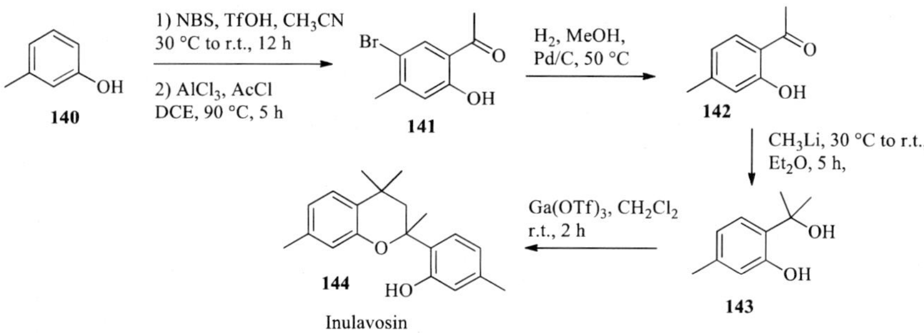

Scheme 29. Synthesis of Inulavosin.

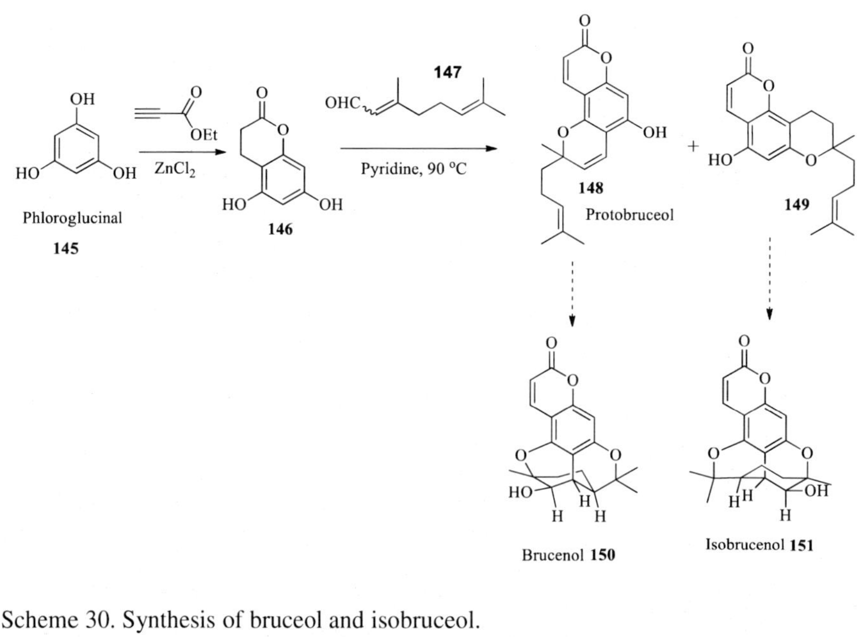

Scheme 30. Synthesis of bruceol and isobruceol.

J. H. George and coworkers described a biomimetic synthesis of bruceol **150** and isobruceol **151**; both are pentacyclic meroterpenoid isolated from *Philotheca brucei* (*Scheme 30*). The key step in the synthesis of both bruceol and isobruceol involved an intramolecular hetero Diels–Alder reaction of an *o*-quinone methide that was formed by dearomatization of an electron rich chromene [58]. First, the synthesis of 5,7-dihydroxycoumarin (**146**) was achieved by the reaction between phloroglucinol (**145**) and ethylpropiolate in the presence of $ZnCl_2$. Treatment of **146** with aldehyde **147** yielded racemic protobruceol-I **148** and **149**, which were then converted to bruceol **150** and isobruceol **151**, respectively.

In 2019, Campagne and coworkers reported a highly enantioselective intramolecular inverse electron-demand hetero Diels–Alder (IIEDHDA) reaction of citral derivative **152** catalyzed by $Fe(OTf)_3$ in the presence of chiral semicorrin ligand to give bicyclic trans dihydropyran derivatives **153** in good isolated yields and excellent enantioselectivities [59]. The presence of pyridine derivative 2,6-lutidine improves the enantioselectivity to a great extent reaching upto 96% ee. The IIEDHDA dihydropyran products **153** possess a usual bicyclic structure with a pending imidazolyl group which can further offer many possibilities for synthetic post-transformations (*Scheme 31*).

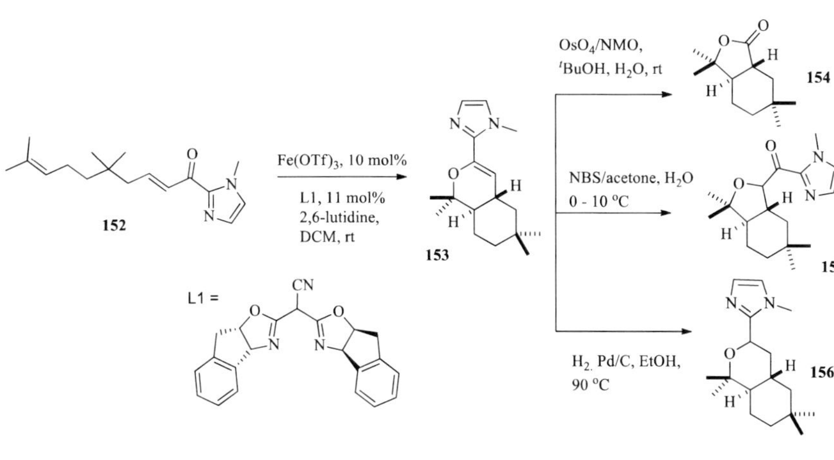

Scheme 31. Synthesis of bicyclic trans dihydropyrans and their applications.

2.4. Oxonium-Ene Reaction

Ene reaction was first discovered by Alder in 1943. It involves an alkene (ene) and an enophile in which the σ bond of allylic hydrogen from alkene suprafacially interacts with the π-bond of the enophile to generate two new σ bonds and a new π-bond [60]. In the ene cyclization reaction, depending upon the enophile used, it can be divided into an oxonium ion ($X=O^+$), iminium ion ($X=N^+$) or thionium ion ($X=S^+$) ene reactions. Hence, it is regarded as one of the versatile method for the synthesis of hetrocyclic compounds (*Scheme 32*) [61].

From a mechanistic perspective, two mechanisms can be proposed for the oxonium-ene cyclization reaction; concerted and stepwise mechanisms. In the concerted mechanism, it proceeds through a cyclic transition state **158** (*Scheme 33*) and in the stepwise mechanism, there is generation of a carbocation intermediate **B**. In the case of concerted mechanism, there is simultaneous elimination of a proton and abstraction of that proton by the lone pair of oxygen as well as cyclization occurred at the same time to give intermediate **158'** which after proton elimination gives final product **159** [62]. In the stepwise mechanism, firstly, the Lewis acid activates the carbonyl oxygen of the aldehyde **160**, which allows nucleophilic attack by the homoallylic alcohol **161** to form an acetal **162**. After the decomposition of the acetal, it generates oxonium ion **A**, which then undergoes cyclization reaction to generate carbocation **B**. The carbocation **B** undergoes elimination of a proton to give the oxonium-ene cyclized products **163**, **164** and **165** [63].

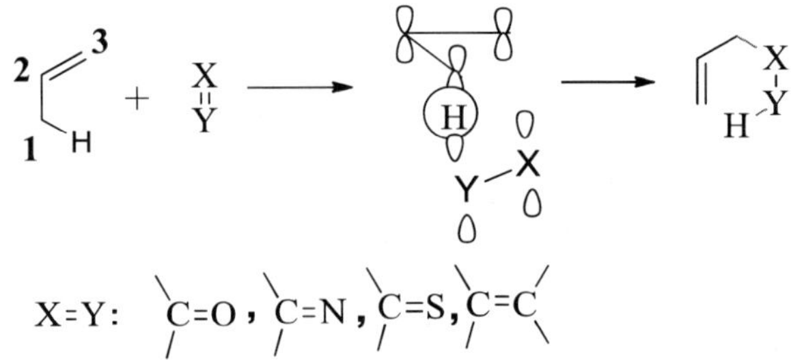

Scheme 32. Ene-reaction.

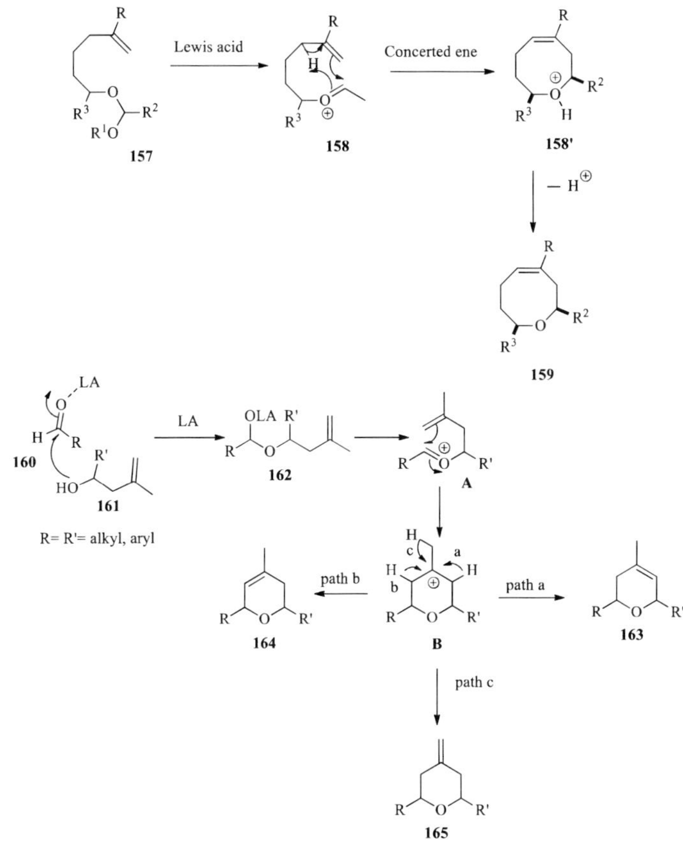

Scheme 33. Concerted vs stepwise mechanisms in oxonium-ene reaction.

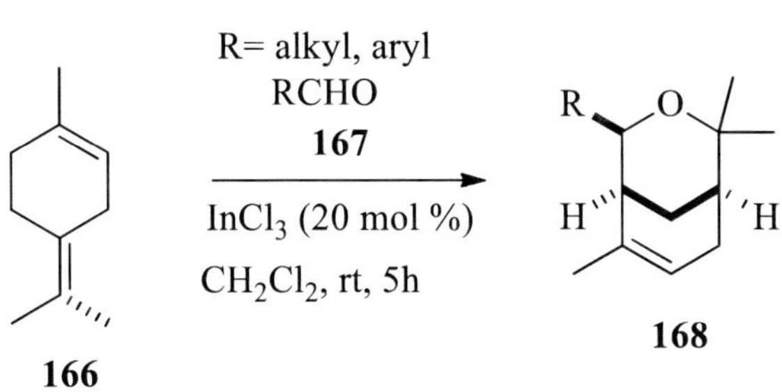

Scheme 34. Synthesis of oxabicyclo [3.3.1] nonenes.

In 2016, Marimuthu and co-workers in 2016, reportedly synthesized oxabicyclo [3.3.1] nonenes **168** *via* (3,5)- oxonium-ene cyclization reaction. Aldehydes **167** and *R*-(+)-limonene **166** under a mild reaction conditions in the presence of catalytic $InCl_3$ in dichloromethane gave corresponding oxabicyclic compound in high yields (*Scheme 34*). The reaction conditions are milder and also consumed shorter duration of time as compared to Saikia and co-workers which was reported earlier [64].

Saikia and coworkers also reported a $Bi(OTf)_3$ catalyzed oxonium-ene cyclization reaction of allenic alcohols **169** with aldehydes **170** for the synthesis of 3,6-dihydro-2*H*-pyrans **171** in good yields. The reaction conditions are very mild and pleasant (*Scheme 35*) [65].

Simultaneously, they were able to synthesized pyranopyrans **172** diastereoselectively, using the same allenic alcohols **169** and aldehydes **170** but in the presence of TMSOTf at – 45 °C. The reaction was achieved *via* oxonium-ene reaction of *β*-allenols and aldehydes followed by [4+2] hetero Diels-Alder type cyclization with another aldehyde molecule (*Scheme 35*). The pyranopyrans **172** can also be directly obtained from the reaction of dihydropyrans **171** and aldehydes **170** under the above reaction conditions.

2.5. Metal Catalyzed Cyclization Reaction

Tetrahydropyran is an important structural skeleton of bioactive natural products. A facile synthesis of *cis*-2,6-disubstituted-3,6-dihydropyrans **174** has been reported in high regio- and stereo-selectivity with high yields. This reaction involved a palladium-catalyzed decarboxylative allylation of 3,4-dihydro-2*H*-pyran derivatives **173** in toluene in the presence of ferrocene derivative adduct. Based on this methodology, the total syntheses of (±)-centrolobine **175** and (±)-decytospolides **A** and **B**, **176A-B** were undertaken as shown in *scheme 36* [66].

Given the allylic feature of enol ethers, the palladium catalyst activated the pyranose system to form salt intermediate **A** which on further decarboxylative coupling produced the C-C bond formation to give compound **174a** (*Scheme 37*).

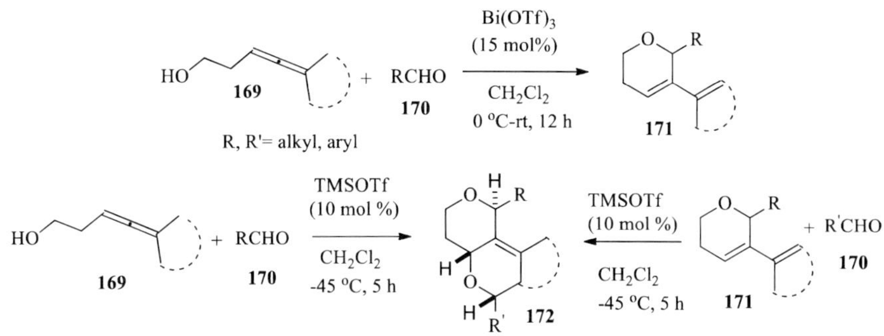

Scheme 35. Synthesis of 3,6-dihydro-2H-pyrans.

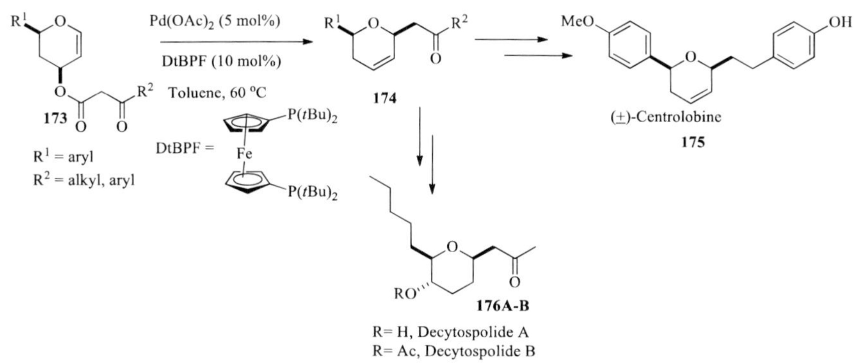

Scheme 36. Synthesis of syntheses of (+)-centrolobine, (+)-decytospolides A and B.

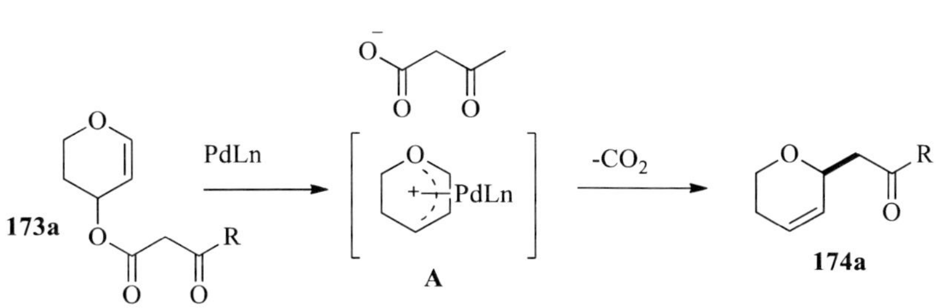

Scheme 37. Mechanism of formation of cis-2,6-disubstituted-3,6-dihydropyrans from 3,4-dihydro-2H-pyran derivatives.

2.6. Miscellaneous

Apart from all the methologies cited above there are few methods which can not be catagorised are discussed in this section. An intramolecular C−C bond formation of oxygen tethered alkyne-epoxides **177** mediated by borontrifluoride etherate solution under ambient conditions was established by Saikia and coworkers. The procedure led to the formation of 3,6- and 5,6-dihydropyrans **178/179** in moderate yields depending upon the substituents R^1 and R^2 on *O*-tethered alkyne epoxides (*Scheme 38*) [67].

Similarly, the same group also synthesized both substituted 5-membered tetrahydrofuran **182** and 6-membered tetrahydropyran derivatives **181** *via* intramolecular hydroalkoxylation of alkenols **180** mediated by borontrifluoride etherate in good yields (*Scheme 39*) [68]. The reaction is highly atom economic and compatible to wide range of functional groups.

They also reported an efficient methodology for the synthesis of tricyclic compound hexahydrobenzo[*de*]isochromanes 1**84** *via* Friedel Crafts and oxa-Pictet-Spengler type reaction from aryl and alkene substituted enol ethers **183**. The reaction was catalyzed by triflic acid and products were obtained in good yields (*Scheme 40*). The compound hexahydrobenzo[*de*]isochromanes **184a/b** are also studied for their antileishmanial activity [69].

Earlier in 2016, they also synthesized isochromans **186** *via* Pictet–Spengler type reaction from substituted enol ethers **185** mediated by trimethylsilyl triflate (TMSOTf) (*Scheme 41*). The methodology was applied for the formal synthesis of key intermediate alcohols **187** for the synthesis of D4 antagonist sonepiprazole (±)-U-101387 **188** and the isochroman (±)-U-54537 [70] **189**.

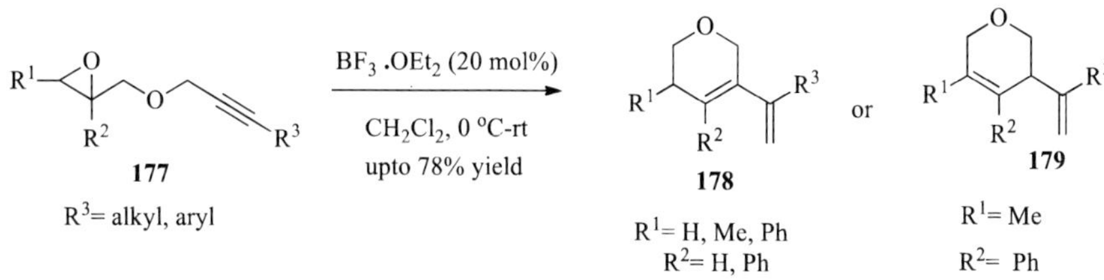

Scheme 38. Synthesis of 3,6- and 5,6-dihydropyrans.

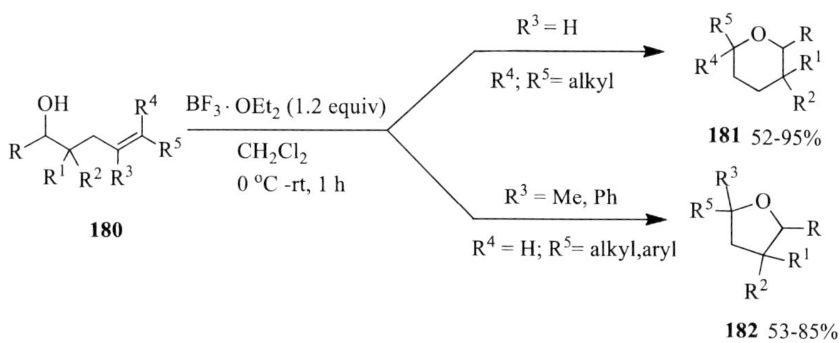

Scheme 39. Syntheis of tetrahydropyrans and furans.

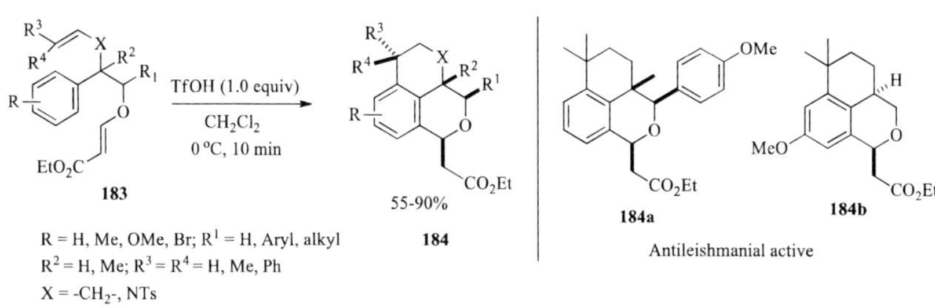

Scheme 40. Synthesis of hexahydrobenzo[de]isochromanes.

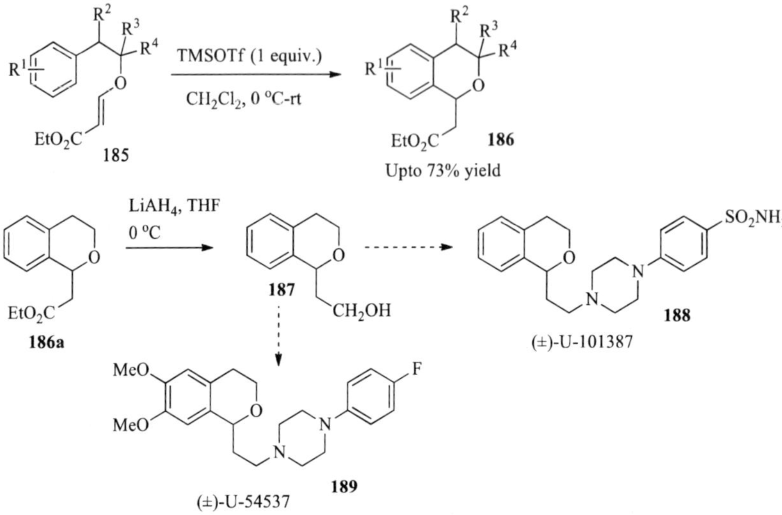

Scheme 41. Synthesis of isochromans and its application towards the synthesis of sonepiprazole (±)-U-101387 and isochroman (±)-U-54537.

REFERENCES

[1] Böhrsch, V. & Blechert, S. (2006). A concise synthesis of (-)-centrolobine via a diastereoselective ring rearrangement metathesis–isomerisation sequence. *Chem. Commun.*, 1968-1970.

[2] Shi, G., Kozlowski, J. F., Schwedler, J. T., Wood, K. V., MacDougal, J. M. & Mclaughlin, J. L. (1996). Muconin and Mucoxin: Additional Nonclassical Bioactive Acetogenins from *Rollinia mucosa*. *J. Org. Chem.*, *61*(XXIII), 7988-7989.

[3] Wender, P. A., Hegde, S. G., Hubbard, R. D. & Zhang, L. (2002). Total Synthesis of (-)-Laulimalide. *J. Am. Chem. Soc.*, *124* (XVIII), 4956-4957.

[4] Nagasawa, T. & Kuwahara, S. (2009). Enantioselective Total Synthesis of Aspergillide C. *Org. Lett.*, *11*(III), 761-764.

[5] (a) Vanderah, D. J., Steadier, P. A., Ciereszko, L. S., Schmitz, F. J., Ekstrand, J. D. & van der Helm, D. (1977). Marine Natural Products. Xenicin: a Diterpenoid Possessing a Nine-Membered Ring from the

Soft Coral, *Xenia elongate*. *J. Am. Chem. Soc*. 5780-5784. (b) Huber, T., Weisheit, L. & Magauer, T. (2015). Synthesis of Xenia diterpenoids and related metabolites isolated from marine organisms. *Beilstein J. Org. Chem.*, *11*, 2521-2539.

[6] Okino, T., Yoshimura, E., Hirota, H. & Fusetani, N. J. (1996). New Antifouling Kalihipyrans from the Marine Sponge *Acanthella cavernosa*. *Nat. Prod.*, *59*(XI), 1081-1083.

[7] Airoldi, C., Mourtas, S., Cardona, F., Zona, C., Sironi, E., D'Orazio, G., Markoutsa, E., Nicotra, F., Antimisiaris, S. G. & La Ferla, B. (2014). Nanoliposomes presenting on surface a cis-glycofused benzopyran compound display binding affinity and aggregation inhibition ability towards Amyloid β1-42 peptide. *Eur. J. Med. Chem.*, *85*, 43-50.

[8] Manriquez, C. P., Souto, M. L., Gavin, J. A., Norte, M. & Fernandez J. J. (2001). Several new squalene-derived triterpenes from *Laurencia*. *Tetrahedron*, *57* (XV), 3117-3123.

[9] Wang, X. J., Zhang, J., Wang, J. D., Qian, P. T., Liu, C. X. & Xiang, W. S. (2013). Novel cyclopentenone derivatives produced by a rare actinobacterial strain *Actinoalloteichus nanshanensis sp*. nov. NEAU 119. *Nat. Prod. Res.*, *27*(XX), 1863-1869.

[10] (a) Kinashi, H., Otake, N., Yonehara, H., Sato, S. & Saito, Y. (1973). The structure of salinomycin, a new member of the polyether antibiotics. *Tetrahedron Lett.*, *14*(XLIX), 4955-4958.

[11] Venkatesham, A., Srinivasa Rao, R., Nagaiah, K., Yadav, J. S., RoopaJones, G. S., Basha, J., Sridhard, B. & Addlagatta, A. (2012). Synthesis of new chromeno-annulated cis-fused pyrano[3,4-*c*]pyran derivatives *via* domino Knoevenagel–hetero-Diels–Alder reactions and their biological evaluation towards antiproliferative activity. *Med. Chem. Commun.*, *3*, 652-658.

[12] McLaughlin, J. L. (2008). Paw Paw and Cancer: Annonaceous Acetogenins from Discovery to Commercial Products. *J. Nat. Prod.*, *71*(VII), 1311-1321. (b) Alali, F. Q., Rogers, L., Zhang, Y. & McLaughlin, J. L. (1998). Unusual bioactive annonaceous

acetogenins from *Goniothalamus giganteus*. *Tetrahedron*, *54*(XXII), 5833-5844.

[13] Washio, T., Nambu, H., Anada, M. & Hashimoto, S. (2007). Catalytic asymmetric hetero-Diels–Alder route to a key intermediate for the synthesis of calyxin L. *Tetrahedron: Asymmetry*, *18*(XXI), 2606-2612.

[14] Pettit, G. R., Herald, C. L., Doubek, D. L. & Herald, D. L. (1982). Isolation and structure of bryostatin 1. *J. Am. Chem. Soc.*, *104*(XXIV), 6846-6848.

[15] Lee, E., Jeong, E. J., Kang, E. J., Sung, L. T. & Hong, H. K. (2001). Total Synthesis of Pamamycin-607. *J. Am. Chem. Soc.*, *123*(XLI), 10131-10132.

[16] Whang, K., Venkataramam, H., Kim, Y. G. & Cha, J. K. (1991). Total synthesis of (-)-citreoviridin and (+)-citreoviral. *J. Org. Chem.*, *56*(XXV), 7174-7177.

[17] Asao, T., Bu̇chi, G., Abdel-Kader, M., Chang, S., Wick, E. & Wogan, G. (1965). The structures of aflatoxins B1 and G1. *J. Am. Chem. Soc.*, *87*(IV), 882-886.

[18] Taylor, E. C. & Jagdmann, Jr. G. E. (1980). Chelation-controlled synthesis of (±)-muscarine. *J. Org. Chem.*, *45*(XVI), 3375-3376.

[19] Hu, L., Lu, X. & Deng, L. (2015). Catalytic Enantioselective Peroxidation of α,β-Unsaturated Aldehydes for the Asymmetric Synthesis of Biologically Important Chiral Endoperoxides. *J. Am. Chem. Soc.*, *137*(XXVI), 8400-8403

[20] Fall, D., Duval, R. A., Gleye, C., Laurens, A. & Hocquemiller, R. (2004). Chamuvarinin, an Acetogenin Bearing a Tetrahydropyran Ring from the Roots of Uvaria chamae. *J. Nat. Prod.*, *67*(VI), 1041-1043.

[21] Lee, J. Y. & Kim, B. H. (1996). Total synthesis of nonactin. *Tetrahedron*, *52*(II), 571-588.

[22] (a) Lu, Z., Lin, Z., Wang, W., Du, L., Zhu, T., Fang, Y., Gu, Q. & Zhu, W. (2008). Citrinin dimers from the halotolerant fungus Penicillium citrinum B-57. *J. Nat. Prod.*, *71*(IV), 543-546. (b) Xu, J., Kjer, J., Sendker, J., Wray, V., Guan, H., Edrada, R., Muller, W. E.

G., Bayer, M., Lin, W., Wu, J. & Proksch, P. (2009). Cytosporones, coumarins, and an alkaloid from the endophytic fungus *Pestalotiopsis sp.* isolated from the Chinese mangrove plant Rhizophora mucronata. *Bioorg. Med. Chem.*, *17*(XX), 7362-7367.

[23] Muramatsu, W. & Nakano, K. (2015). Efficient C(sp^3)-H Bond Functionalization of Isochroman by AZADOL Catalysis. *Org. Lett.*, *17*(VI), 1549-1552.

[24] (a) Dobson, T. A. & Humber, L. G. (1975). Pyrano heterocycles. I. Syntheses of isochromans and the novel thieno[3,2-*c*]pyran, benzothieno[3,2-*c*]pyran, benzothieno[2,3-*c*]pyran, and pyrano[4,3-*b*] benzofuran systems. *J. Heterocyclic Chem.*, *12*(III), 591-594. (b) Yamato, M., Hashigaki, K., Ishikawa, S., Kokubu, N., Inoue, Y., Tsuruo, T. & Tashiro, T. (1985). Synthesis and antitumor activity of tropolone derivatives. *J. Med. Chem.*, *28*(VIII), 1026-1031.

[25] Richter, H., Rohlmann. R. & Mancheño, O, G. (2011). Catalyzed Selective Direct α and γ Alkylation of Aldehydes with Cyclic Benzyl Ethers by Using T(+)BF4- in the Presence of an Inexpensive Organic Acid or Anhydride. *Chem. Eur. J.*, *17*(XLI), 11622-11627.

[26] Kitagawa, I., Zhang, R., Hori, K., Tsuchiya, K. & Shibuya, H. (1992). Indonesian medicinal plants. II. Chemical structures of pongapinones A and B, two new phenylpropanoids from the bark of *Pongamia pinnata* (Papilionaceae). *Chem. Pharm. Bull.*, *40*(VIII), 2041-2043.

[27] (a) Gerha¨user, C., Lee, S. K., Kosmeder, J. W., Moriarty, R. M., Hamel, E., Mehta, R. G., Moon, R. C. & Pezzuto, J. M. (1997). Regulation of ornithine decarboxylase induction by deguelin, a natural product cancer chemopreventive agent. *Cancer Res.*, *57*(XVI), 3429-3435 (b) Fang, N. & Casida, J. E. (1999). New bioactive flavonoids and stilbenes in Cube resin Insecticide. *J. Nat. Prod.*, *62*(II), 205-210. (c) Fang, N. & Casida, J. E. (1998). Anticancer action of cube insecticide: correlation for rotenoid constituents between inhibition of NADH: ubiquinone oxidoreductase and induced ornithine decarboxylase activities. *Proc. Natl. Acad. Sci. U.S.A.*, *95*(VII), 3380-3384.

[28] (a) Prins, H. J. (1919). *Proc. Acad. Sci. Amsterdam*, *22*, 51. (b) Prins, H. J. (1919). The reciprocal condensation of unsaturated organic compounds. *Chem. Weekbl.*, *16*, 1510-1526.

[29] Hanschke, E. (1955). The Prins reaction. III. The reaction of allylcarbinol with aldehydes and ketones. *Chem. Ber.*, *88*, 1053-1061.

[30] (a) Arundale, E. & Mikesa, L. A. (1952). The olefin-aldehyde condensation. The Prins reaction. *Chem. Rev.*, *51*, 505-555. (b) Adams, D. R. & Bhatnagar, S. P. (1977). The Prins reaction. *Synthesis*, *10*, 661-672. (c) Alder, R. W., Harvey, J. N. & Oakley, M. T. (2002). Aromatic 4-Tetrahydropyranyl and 4-Quinuclidinyl Cations. Linking Prins with Cope and Grob. *J. Am. Chem. Soc.*, *124*(XVIII), 4960-4961.

[31] Yadav, J. S., Subba Reddy, U. V. & Subba Reddy, B. V. (2014). Stereoselective synthesis of tetrahydropyran-tetrahydrofuran (THP-THF) core of (+)-muconin *via* Prins cyclization. *Tetrahedron Lett.*, *55*(XXIX), 3860-3863.

[32] Zhang, W., Xiao D. & Wang, B. (2018). A concise total synthesis of cochlearoid B. *Org. Biomol. Chem.*, *16*(XVIII), 3358-3361.

[33] (a) Zhou, F. J., Nian, Y., Yan, Y., Gong, Y., Luo, Q., Zhang, Y., Hou, B., Zuo, Z. L., Wang, S. M., Jiang, H. H., Yang J. & Cheng, Y. X. (2015). Two new classes of T-type calcium channel inhibitors with new chemical scaffolds from Ganoderma cochlear. *Org. Lett.*, *17*(XII), 3082-3085. (b) Paterson, R. R. (2006). Ganoderma - A therapeutic fungal biofactory. *Phytochemistry*, *67*(XVIII), 1985-2001.

[34] Chennaiah, A., Dubbu, S., Parasuraman, K. & Vankar, Y. D. (2018). Stereoselective Synthesis of 1,2-Annulated Sugars Having Substituted Tetrahydropyran/(-furan) Scaffolds Using the Prins-Reaction. *Eur. J. Org. Chem.*, (47), 6706-6713.

[35] Kantee, K., Rukachaisirikul, V. & Tadpetch, K. (2016). Synthesis of tetrahydropyranyl diarylheptanoids from *Dioscorea villosa*. *Tetrahedron Lett.*, *57*(XXXI), 3505-3509.

[36] Lv, H. & She, G. (2012). Naturally occurring diarylheptanoids - a supplementary version. *Rec. Nat. Prod.*, *6*(IV), 321-333.

[37] (a) Dong, S. H., Nikolic´, D., Simmler, C., Qiu, F., van Breemen, R. B., Soejarto, D. D., Pauli, G. F. & Chen, S. N. (2012). Diarylheptanoids from *Dioscorea villosa* (Wild Yam). *J. Nat. Prod.*, *75*(XII), 2168-2177. (b) Woo, K. W., Moon, E., Kwon, O. W., Lee, S. O., Kim, S. Y., Choi, S. Z., Son, M. W. & Lee, K. R. (2013). Anti-neuroinflammatory diarylheptanoids from the rhizomes of *Dioscorea nipponica*. *Bioorg. Med. Chem. Lett.*, *23*(XIII), 3806-3809.

[38] Subba Reddy, B. V., Reddy, M. R., Reddy, S. G., Sridhar, B. & Kumar, S. K. (2015). Tandem Prins Cyclization for the Stereoselective Synthesis of the 4,5-Diaryl-hexahydropyrano[3,4-*c*] chromene Skeleton of Calyxins I and J. *Eur. J. Org. Chem.*, (XIV), 3103-3108.

[39] Han, Y., Tang, X., Cheng, J. & Ma, S. (2016). Indium(III) Triflate-Catalyzed Efficient Prins-Type Cyclization of *β*-Allenols and Aldehydes. *Adv. Synth. Catal.*, *358*(XXIV), 4019-4029.

[40] Scoccia, J., Pérez, S. J., Sinka, V., Cruz, D. A., López-Soria, J. M., Fernández, I., Martín, V. S., Miranda, P. O. & Padrón, J. I. (2017). Direct Access to 2,3,4,6-Tetrasubstituted Tetrahydro-2*H*-pyrans *via* Tandem S_N2'-Prins Cyclization. *Org. Lett.*, *19*(XVIII), 4834-4837.

[41] Saikia, A. K., Ghosh, P. & Kautarya, A. K. (2016). Synthesis of 4-trifluoromethanesulfonate substituted 3,6-dihydropyrans and their application in various C-C coupling reactions. *RSC Adv.*, *6*(L), 44774-44781.

[42] Liu, L., Kaib, P. S. J., Tap, A. & List, B. (2016). A General Catalytic Asymmetric Prins Cyclization. *J. Am. Chem. Soc.*, 138(XXXIV), 10822-10825.

[43] Shit, S., Devi, N., Devi, N. R. & Saikia, A. K. (2019). Stereoselective synthesis of hexahydrofuro[3,4-*b*]furan-4-ol and its dimer *via* tandem Prins and pinacol rearrangement. *Org. Biomol. Chem.*, *17*(IX), 7398-7407.

[44] Gogoi, P., Das, V. K. & Saikia, A. K. (2014). Diastereoselective Synthesis of Substituted Tetrahydrofurans *via* Prins Cyclization of Enol Ethers. *J. Org. Chem.*, *79*(XVIII), 8592-8598.

[45] Borthakur, U., Biswas, S. & Saikia, A. K. (2019). Vinylsilanes in Highly Diastereo- and Regio-Selective Synthesis of Dihydropyrans. *ACS Omega*, *4*(II), 2630-2636.

[46] Sarkar, S., Devi, N., Porashar, B., Ruidas, S. & Saikia, A. K. (2019). Stereoselective Synthesis of 4-*O*-Tosyltetrahydropyrans *via* Prins Cyclization Reaction of Enol Ethers. *SynOpen*, *3*(I), 36-45.

[47] Aubert, C., Buisine, O. & Malacria, M. (2002). The Behavior of 1,*n*-Enynes in the Presence of Transition Metals. *Chem. Rev.*, *102*(III), 813-834.

[48] Ibe, K., Yamada, T. & Okamoto, S. (2019). Synthesis and vitamin D receptor affinity of 16-oxa vitamin D3 analogues. *Org. Biomol. Chem.*, *17*(XLVIII), 10188-10200.

[49] Johnson, T. C., Chin, M. R., Han, T., Shen, J. P. Rana, T. & Siegel, D. (2016). *J. Am. Chem. Soc.*, *138*(XVIII), 6068-6073.

[50] Lin, A., Zhang, Z. W. & Yang, J. (2014). Iron-Catalyzed Reductive Cyclization of 1,6-Enynes. *Org. Lett.*, *16*(II), 386-389.

[51] Gresham, T. L. & Steadman, T. R. (1949). Diels-Alder type reaction with formaldehyde. *J. Am. Chem. Soc.*, *71*, 737-738.

[52] Murray, L. A. M., Fallon, T., Sumby, C. J. & George, J. H. (2019). Total Synthesis of Naphterpin and Marinone Natural Products. *Org. Lett.*, *21*(XX), 8312-8315.

[53] Lam, H. C., Spence, J. T. J. & George, J. H. (2016). Biomimetic Total Synthesis of Hyperjapones A-E and Hyperjaponols A and C. *Angew. Chem. Int. Ed.*, *55*(XXXV), 10368-10371.

[54] Gutiérrez, R. M. P., Mitchell, S. & Solis, R. V. (2008). Psidium guajava: A review of its traditional uses, phytochemistry and pharmacology. *J. Ethnopharmacol.*, *117*(I), 1-27.

[55] Ning, S., Liu, Z., Wang, Z., Liao, M. & Xie, Z. (2019). Biomimetic Synthesis of Psiguajdianone: Guided Discovery of the Meroterpenoids from Psidium guajava. *Org. Lett.*, *21*(XXI), 8700-8704.

[56] Kim, T. H., Hatano, T., Okamoto, K., Yoshida, T., Kanzaki, H., Arita, M. & Ito, H. (2017). Antifungal and ichthyotoxic

sesquiterpenoids from Santalum album heartwood. *Molecules.*, *22*(VII), 1139-1142.

[57] Gonga, P. X., Lia, H. J., Wang, M., Cheng, Y. F. & Wua, Y. C. (2019). Synthesis of natural product inulavosin via $Ga(OTf)_3$-Catalyzed Hetero Diels-Alder Dimerization of salicyl alcohol derivative. *Natural Product Research*, *33*(XX), 2911-2916.

[58] Day, A. J., Sumby, C. J. & George, J. H. (2019). Biomimetic Synthetic Studies on the Bruceol Family of Meroterpenoid Natural Products. *J. Org. Chem.* (ahead of print).

[59] Lauberteaux, J., Lebrun, A., Lee, A., Mauduit, M., Marcia de Figueiredo, R. & Campagne, J. M. (2019). Iron-Catalyzed Enantioselective Intramolecular Inverse Electron-Demand Hetero Diels-Alder Reactions: An Access to Bicyclic Dihydropyran Derivatives. *Org. Lett.*, *21*(XXIV), 10007-10012.

[60] (a) Golstein, M. J. & Hoffmann, R. (1971). Symmetry, topology and aromaticity. *J. Am. Chem. Soc.*, *93*(XXIII), 6193-6204. (b) Woodward, R. B. & Hoffmann, R. (1970). The conservation of orbital symmetry Academic Press, New York. (c) Fleming, I. (1976). Frontier orbitals and organic chemical reactions Wiley, London. (d) Saha, P. & Saikia, A. K. (2018). Ene cyclization reaction in heterocycle synthesis. *Org. Biomol. Chem.*, *16*(XIV), 2820-2840.

[61] (a) Mikami, K., Sawa, E. & Terada, M. (1991). Asymmetric catalysis by chiral titanium perchlorate for carbonyl-ene cyclization. *Tetrahedron: Assymmetry*, *2*(XIX), 1403-1412. (b) Ohmura, H., Smyth, G. D. & Mikami, K. (1999). The First Example of (2,5) Ene Cyclization: Solid Acid-Catalyzed Oxonium-Ene Reaction. *J. Org. Chem.*, *64*(XVI), 6056-6059.

[62] Blumenkopf, T. A., Look, G. C. & Overman, L. E. (1990). Formation of Δ4-oxocenes from Lewis acid promoted cyclizations of 5-hexenyl acetals. Evidence for a concerted ene cyclization mechanism. *J. Am. Chem. Soc.*, *112*(XI), 4399-4403.

[63] Bondalapati, S., Reddy, U. C., Saha, P. & Saikia, A. K. (2011). An efficient synthesis of dihydro- and tetrahydropyrans *via* oxonium-ene cyclization reaction. *Org. Biomol. Chem.*, *9*(IX), 3428-3438.

[64] Singaram, K. & Marimuthu, D. (2016). Indium(III) chloride-catalyzed facile synthesis of oxabicyclo[3.3.1]nonenes *via* (3,5)-oxonium-ene cyclization. *Syn Commun.*, *46*(XV), 1257-1262.

[65] Devi, N. R., Sultana, S., Borah, M. & Saikia, A. K. (2018). Regio- and Diastereoselective Synthesis of Dihydropyrans and Pyranopyrans *via* Oxonium-Ene Reaction of β-Allenols and Aldehydes. *J. Org. Chem.*, *83*(XXIV), 14987-14998.

[66] Zeng, J., Tan, Y. J., Ma, J., Leow, M. L., Tirtorahardjo, D. & Liu, X. W. (2014). Facile Access to cis-2,6-Disubstituted Tetrahydropyrans by Palladium-Catalyzed Decarboxylative Allylation: Total Syntheses of (±)-Centrolobine and (+)-Decytospolides A and B. *Chem. Eur. J.*, *20*(II), 405-409.

[67] Ghosh, P., Saha, P., Bondalapati, S., Indukuri, K. & Saikia, A. K. (2014). Lewis Acid Mediated Intramolecular C-C Bond Formation of Alkyne-Epoxide Leading to Six-Membered Nitrogen and Oxygen Heterocycles. *J. Org. Chem.*, *79*(IX), 4119-4124.

[68] Sultana, S., Devi, N. R. & Saikia, A. K. (2015). Synthesis of Substituted Tetrahydropyrans and Tetrahydrofurans *via* Intramolecular Hydroalkoxylation of Alkenols. *Asian J. Org. Chem.*, *4*(XI), 1281-1288.

[69] Saikia, A. K., Sultana, S., Devi, N. R., Deka, M. J., Tiwari, K. & Dubey, V. K. (2016). Diastereoselective synthesis of substituted hexahydrobenzo[*de*]isochromanes and evaluation of their antileishmanial activity. *Org. Biomol. Chem.*, *14*(III), 970-979.

[70] Sarkar, S., Sultana, S., Indukuri, K., Unnava, R. & Saikia, A. K. (2016). Synthesis of Isochroman Derivatives via Oxa-Pictet–Spengler Reaction of Vinylogous Esters. *Synthesis*, *48*(XI), 1727-1733.

In: Heterocycles
Editor: Jonas M. Danielsen
ISBN: 978-1-53617-804-3

Chapter 3

ORIGIN OF PUCKERING (BUCKLING) OF PLANAR HETEROCYCLES AND METHODS OF ITS SUPPRESSION

N. N. Gorinchoy[1] and I. B. Bersuker[2]
[1]Institute of Chemistry, Academy of Science of Moldova, Moldova Republic
[2]Department of Chemistry, The University of Texas at Austin, Austin TX, US

ABSTRACT

This chapter is devoted to one of the most important structural features of heterocyclic compounds that defines their properties, the deviation from the planar configuration in the form of puckering or buckling, and to the methods of its suppression. First, we show that all the distortions of the planar configuration have a common origin, the pseudo Jahn-Teller effect (PJTE), and we explain how this effect produces the atomic out-of-plane displacements in function of the systems' parameters. The latter include the ground and low-lying excited states of the undistorted configuration, the energy gaps between them, and the vibronic coupling constants. We demonstrate how these PJTE parameters can be evaluated by means of electronic structure calculations, using some hexa-heterocyclic compounds as examples. The knowledge of the main features and explicit values of the parameters of the PJTE open a way to directly influence this effect and its produced distortions of the planar configuration by means of specific, *targeted* external implications (perturbations) in the form of oxidation, reduction, excitation, chemical substitution, and coordination to outer atoms or atomic groups. All these methods were successfully realized. Illustrative examples include a series of carbon and silicon tetra- and hexa-cyclic compounds, and more complicated tricyclic and polycyclic systems like dithiine containing dithianon, thiantrene, anthracene, sumanenes, etc. as well as extended 2D systems including silicene, phosphorene, germanene, and graphitic carbon nitride.

1. INTRODUCTION

Heterocyclic compounds have enumerable, unlimited applications in physics, chemistry, biology, and materials science due to their specific structural and electronic properties, and their importance was appended more recently by applications of their extended two-dimensional (2D) graphene-like forms as materials for novel technologies [1, 2]. This chapter is devoted to the special property of heterocyclic systems, the deviation of their structure from the planar configuration in the form of puckering or buckling. These out-of-plane distortions influence all the properties of the compound, notably the chemical reactivity, and the band gap and charge mobility in the extended 2D systems. As stated by a group of experts [3], "the native buckling can have dramatic implications in the overall physics and chemistry" of 2D or pseudo 2D systems, and "potentially serve as a nanotechnology platform…for fundamental research and applications as reconfigurable devices for electronic, spintronic, photonic, thermal, energy, mechanical, chemical, and sensor nanosystems".

The planarity and deviation from planarity of heterocyclic systems is thus one of the most important, sometimes crucial, properties in defining their properties. This raises the problem of influencing (manipulating) the planarity of these systems by means of external implications (perturbations). Some authors suggest influencing the structural features of the systems by changing the chemical composition, other possibilities were also suggested [2, 4]. Many attempts to influence the planarity of cyclic systems employ the considerations of atomic orbital hybridization. Indeed, compared with the ideally flat graphene, the interatomic distance in the series of buckled hexagonal honeycomb quasi-flat structures of Si, Ge, and Sn increases significantly from left to right (~1.42 Å in graphene, ~2.28 Å in silicene, etc.). This weakens the overlap between their atomic p_z orbitals and reduces the stabilizing contribution of π bonding in the flat configuration, making the atomic sp^3 hybridization and the consequent buckled (honeycomb) structure advantageous. This qualitatively correct consideration of the influence of atomic orbital hybridization on local structural elements serves as an illustrative feature of the distortion effect in some particular cases, but it is insufficient to explain the origin of puckering or buckling in more complicated situations, which cannot be reduced to just hybridizations.

On the other hand, it was shown that all deviations from high-symmetry configurations (in our case the planar configurations) of polyatomic systems in nondegenerate states, in our case in the form of puckering or buckling, are due to, and only to the pseudo Jahn-Teller effect (PJTE) [5, 6]. In application to a series of cyclic and heterocyclic compounds the PJTE explains the origin of their out-of-plan distortions (see [6] and references therein; puckering as a PJTE was first formulated in [7]). The knowledge of the PJTE origin and mechanism of out-of-plane distortions of 2D systems leads directly to the possibility to change their planarity by means of external influence that changes the parameters of the PJTE. In the simplified two-level formulation (see below), the condition of instability of the high-symmetry configuration of the system is a relation between its main three parameters [5, 6],

$$\Delta < F^2/K_0$$

where Δ is the energy gap to the excited state with appropriate symmetry (active in the PJTE), F is the constant of vibronic coupling to the latter, and K_0 is the primary force constant characterizing the stiffness of the system with respect to the distortion. Following this PJTE instability condition, one can control (manipulate) the structural instability and transformations of molecular systems and solids by means of *targeted* electronic rearrangements: charge transfer, excitation, ionization, redox processes, partial molecular orbital (MO) population changes by coordination, intermolecular (van-der-Waals or ionic-polarization) interactions, external pressure, electromagnetic fields, as well as appropriate chemical substitutions.

The procedure of influencing the planarity of heterocyclic systems, described below, starts with electronic structure calculations of the ground and low-lying excited states in the high-symmetry (planar) configuration, revealing the active excited states which produce the instability of the ground state, their energy, symmetry, MO origin, and the numerical values of the parameters in the condition of instability (1). This gives us the necessary insight on how we can influence (enhance or suppress) the condition of instability of the planar configuration, meaning manipulate the puckering or buckling of the cycling system with external influence mentioned above. In principle, there are several possibilities to influence the PJTE distortions, among which the following seem to be most practical for heterocycles: (1) increasing the energy gap Δ between the electronic states (that are mixed by the PJT vibronic coupling) by means of weak coordination with outside atoms or atomic groups (see below), (2) changing the positions and symmetry of the PJTE active ground and excited states via a redox process, by removing electrons from, or adding electrons to specific electronic states, and (3) modifying the system by means of substitutions that do not violate the cyclic quasi-2D character of the system, but change the PJTE parameters.

All these possibilities were effectively realized in changing the planarity of molecular systems [8, 9], reviewed below in this chapter. We briefly introduce the PJTE and show how it can be employed to control the puckering or buckling of planar systems using the already published papers on this topic as illustrative examples. In Section 2 we introduce the PJTE. Section 3 shows briefly the technical procedure in application of the PJTE theory to the specific problem of the origin of instability of planar configurations of molecular systems. Section 4 reviews the publications in which the methods of suppressing the PJTE in puckered (buckled) molecular and some extended 2D systems, and thus restoring their planar configuration, are realized. Concluding remarks are given in Section 5.

2. The Pseudo Jahn-Teller Effect. Basic Formulations

The equilibrium geometry of any polyatomic system is defined as corresponding to the point of the minimum of the adiabatic potential energy surface (APES), where its first derivatives are zero and the second derivatives are positive. This stable configuration does not necessarily coincide with the highest-symmetry or any high-symmetry configuration. We define the latter as the reference configuration corresponding to the point of the APES where the first derivatives are zero too, but the second derivative in some direction may be either positive or negative, meaning the high-symmetry configuration may be unstable. Denote the energy of the system as a function of normal displacements Q_Γ by $E(Q_\Gamma)$ (the index Γ stands for the symmetry representation of the group of symmetry of the system, and the symmetrized nuclear Q_Γ displacements are well defined by the group theory [5]). At the point of minimum of the APES at $Q_\Gamma=0$ the curvature K of $E(Q_\Gamma)$ in the Q_Γ direction,

$$K_\Gamma = (\partial^2 E/\partial Q_\Gamma^2)_0 \qquad (1)$$

is positive, $K_\Gamma > 0$. Very often the geometry of the system at this point of equilibrium on the APES does not coincide with the highest possible (or even with any high) symmetry expected from general symmetry considerations. For instance, linear molecules are bent at equilibrium, planar molecules are puckered, octahedral complexes are elongated, or compressed or tilted, cubic crystals are tetragonal or trigonal distorted (or have several structural phases), etc. The PJTE is the general driving force for all these distortions of high-symmetry configurations in non-degenerate states [5, 6].

If at the point of high-symmetry configuration on the APES the system is structurally unstable with respect to some nuclear displacements Q_Γ, than $K_\Gamma < 0$ in this direction. Let us analyze this possibility. The general formula for the energy is $E = \langle\psi_0|H|\psi_0\rangle$, where H is the Hamiltonian of the and ψ_0 is the wavefunction of the nondegenerate ground state. Substituting E in Eq. (1), we get (omitting the Γ index for simplicity):

$$K = \langle\psi_0|(\partial^2H/\partial Q^2)_0|\psi_0\rangle + 2\langle\psi_0|(\partial H/\partial Q)_0|\psi_0'\rangle \tag{2}$$

where $\psi_0' = (\partial\psi_0/\partial Q)_0$, and all the functions in Eq. (2) are now taken at the point of high-symmetry configuration. Denoting the first term in (2) by K_0,

$$K_0 = \langle\psi_0|(\partial^2H/\partial Q^2)_0|\psi_0\rangle \tag{3}$$

and the second one as K_v,

$$K = K_0 + K_v \tag{4}$$

we notice that $K_v < 0$. Indeed, presenting the Hamiltonian of the system as a sum,

$$H = H_0 + \sum_\Gamma \left(\frac{dH}{dQ_\Gamma}\right)_0 Q_\Gamma + \ldots \tag{5}$$

where H_0 is the Hamiltonian of the system with fixed nuclei in the high-symmetry (planar) configuration, while the second and next terms stand for the coupling with the nuclear displacements (vibronic coupling), and presenting ψ_0' in terms of the second order perturbation theory with respect to the vibronic coupling, we get directly:

$$K_v = -2\sum_n \frac{\left|\langle\Psi_0|(\partial H/\partial Q_\alpha)_0|\Psi_n\rangle\right|^2}{E_n - E_0} \tag{6}$$

where ψ_n are the wavefunctions of the excited states and E_n are their energies. This K_v term thus contributes to the instability of the system.

Therefore, if $K_0 > 0$, the K_v contribution is the only source of instability. Note that, as mentioned above, we define the reference high-symmetry configuration with respect to the displacements Q as the point of the APES at which $(\partial E/\partial Q)_0 = 0$, meaning zero distortion force in this direction. The matrix elements in Eq. (6) are off-diagonal vibronic coupling constants [5],

$$F_{0n} = \langle\psi_0|(\partial H/\partial Q)_0|\psi_n\rangle \tag{7}$$

that mix the ground and excited states under the nuclear displacements Q; therefore K_v is termed vibronic contribution. Together with the K_0 value and the energy gap $2\Delta_{0n} = E_n - E_0$ between the mixing states, F_{0n} are the main parameters of the PJTE.

In a series of papers, beginning in 1980 (see in [5, 6]), it was proved that for any polyatomic system in the high-symmetry configuration

$$K_0 > 0 \tag{8}$$

and hence the vibronic contribution K_v is the only source of instability of any polyatomic system in nondegenerate states. This is the theorem of instability of high-symmetry configurations. If $K_0 > 0$ for the high-symmetry configuration of any polyatomic system, then a negative curvature $K = (K_0 + K_v) < 0$ can be achieved only due to the negative vibronic coupling component K_v, and only if $|K_v| > K_0$. It follows that any distortion of the high-symmetry configuration is due to, and only to the mixing of its ground state with excited electronic states by the distortive nuclear displacements realized via the vibronic coupling terms in Eq. (6). It means that to rationalize the origin of molecular shapes (or crystal structures) by reasonably starting with some high-symmetry (force-equilibrated) configuration, we notice that there is no way to get this configuration distorted unless we involve its excited electronic states via the vibronic coupling. The latter softens the system with respect to certain nuclear displacements ($K_v < 0$), and if this softening is larger than the original (non-vibronic) stiffness K_0 in this direction, the system becomes unstable with respect to the distortions under consideration, leading to its equilibrium geometry of lower symmetry, or to dissociation.

As nuclear displacements lay in the base of the formation, deformation, and transformation of polyatomic systems, there should be at least one excited state involved (along the ground state) in these processes or, if we consider these properties of the system in the excited state, another excited state or the ground state should be involved. It follows that at least two electronic states of the reference configuration should be involved in rationalization of structural changes of any polyatomic system. This is the two-electronic-state paradigm in formation, deformation, and transformation of molecular systems and solids, or compressed, two electronic states in transformations (TEST) paradigm [6].

Thus starting with the planar configuration of a polyatomic system and looking for the reason why it may be unstable with respect to out-of-plan distortions (lower-symmetry normal nuclear displacements Q), we come to the negative contributions of the excited states K_v to the curvature (stability) of the APES in the ground state K_0, given by Eq. (6). It contains a theoretically infinite number of terms including all the corresponding excited states that are mixed with the ground one under the nuclear displacements Q, but practically this number is limited, first by symmetry, and second by energy gaps. Indeed, the matrix elements in the nominators of Eq. (6), the mentioned above vibronic coupling constants $F_{0n} = \langle\psi_0|(\partial H/\partial Q)_0|\psi_n\rangle$ are non-zero only if the product of symmetries of the two wavefunctions of the ground and excited states contains the symmetry of the symmetrized (normal) displacement Q, and the contribution of the higher excited states is strongly diminished by the increasing energy gaps $2\Delta_{0n} = E_n - E_0$ in the denominators. As Q is a low-symmetry displacement, the two wavefunction should have different symmetry. The meaning of the vibronic coupling constant F is that it reflects the rate of increase of the overlap of the two wavefunctions (under the Q displacements) which is zero in the high-symmetry configuration.

For these reasons only one, two, or several excited states effectively take part in the softening of the high-symmetry configuration of the ground state, and they can be revealed in a two-level (or n-level) problem. The active excited state discriminates also the distortion coordinate Q (for which $F_{0n} \neq 0$) from several possible in the high-symmetry configuration. Along this coordinate the contribution of all the other terms in Eq. (6) can be taken into account as a small correction.

In the simplest case of the PJTE when only one nondegenerate excited state contributes significantly to the instability of the ground state in the planar configuration, we have two interacting electronic states ψ_{01} and ψ_{02}, with an energy gap between them E_{02}- E_{01} = 2Δ. Then, by taking the energy read-off in the middle of this gap, and including the linear vibronic coupling term of the Hamiltonian (5) as a perturbation, we get a secular equation for the energy corrections as follows:

$$\begin{vmatrix} -\Delta - \varepsilon & F_\Gamma Q_\Gamma \\ F_\Gamma Q_\Gamma & \Delta - \varepsilon \end{vmatrix} = 0 \quad (9)$$

with the roots

$$\varepsilon_{1,2} = \pm [\Delta^2 + (F_\Gamma Q_\Gamma)^2]^{1/2} \quad (10)$$

which by adding the elastic energy terms $(1/2)K_{01,2}^{\Gamma}(Q_\Gamma)^2$ yields the following cross section of the APES along the Q_Γ coordinate, illustrated in Figure1:

$$E_{1,2}(Q_\Gamma) = (1/2)K_{01,2}^{\Gamma}(Q_\Gamma)^2 \pm [\Delta^2 + (F_\Gamma Q_\Gamma)^2]^{1/2} \quad (11)$$

These two function's dependence on Q at $Q_\Gamma = 0$ is strongly controlled by the relation between the three parameters: Δ, F_Γ, and K_0^Γ, more precisely, by the relation between the energy half-gap Δ and the ratio F_Γ / K_0^Γ. First, by derivation, we get the curvatures of the two functions (11) at $Q_\Gamma = 0$ as follows:

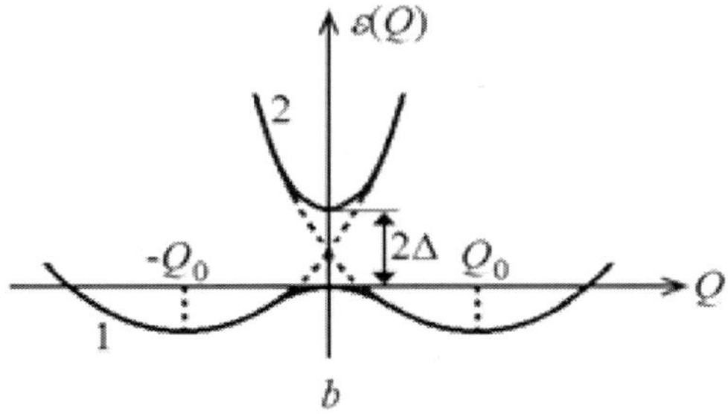

Figure 1. Two branches of the APES produced by the PJTE in a polyatomic system in the two-level approximation with one distortion coordinate Q. Under the condition (14) the ground state is unstable at Q = 0, and it has two equivalent minima positions at Q = ± Q_0.

$$K_1^\Gamma = K_{01}^\Gamma - (F_\Gamma)^2/\Delta \tag{12}$$

$$K_2^\Gamma = K_{02}^\Gamma + (F_\Gamma)^2/\Delta \tag{13}$$

We see that, as a result of the vibronic coupling between the two states, the curvature of the APES along Q_Γ of the lower state decreases, while that of the higher state increases by the same amount (Figure 1). If

$$\Delta < (F_\Gamma)^2/K_{01}^\Gamma \tag{14}$$

the curvature of the lower state becomes negative, meaning instability in the Q_Γ direction (Figure 1). This is the PJTE in the case of two interacting electronic states [6]. It leads to the softening (always) and instability (under the criterion (14)) of the lower state and hardening (always) of the upper one. It is important to emphasize that in the condition of instability (14), there are no limitation on any of the three parameters, taken separately (notably, the most visible energy gap 2Δ), as their magnitude may be compensated by the combination of the other two of them.

Note that only occupied with unoccupied energy levels interact to produce the PJTE. As seen from Eqs. (12) and (13), if the two interacting states are both occupied, the PJTE destabilization of the lower state is compensated by the equivalent stabilization of the upper state resulting in a total zero effect. If the electron populations of the lower and upper states are different, the PJTE is approximately proportional to their difference. Note also that, as mentioned above, the realization of the PJTE is limited by the condition that the low-symmetry atomic displacements Q produce nonzero vibronic coupling *F* in Eq. (7), meaning sufficiently strong overlap between the wavefunctions, which are orthogonal in the high-symmetry configuration, and this was shown to improve the covalence bonding between the corresponding atoms [5, 6]. In other words, the PJTE is the driving force for atomic rearrangements (distortions) that lower the total energy of the system, mainly by improving the covalence bindings with possible additional contributions from charge redistributions (polarizations) [6]. All the possible (qualitatively visible) reasons of distortions of high-symmetry configurations are particular cases of the PJTE. Steric repulsions too are incorporated in the PJTE: they enter into the condition of PJTE instability (14) via the K_0 parameter, the primary force constant (the stiffness) of the system in the high-symmetry configuration with respect to the distortions under consideration.

3. Revealing the Excited States and PJTE Parameters that Produce the Instability of the Flat Configuration of Heterocyclic Systems

Based on the general proof that all the deviations from high-symmetry configuration of polyatomic systems in nondegenerate states are of PJTE origin, we can find out the main features of the planar configuration of 2D systems that are responsible for the out-of-plan instability by means of electronic structure calculations including vibronic coupling. In view of further applications (manipulations), the PJTE parameters of interest include the most active excited states responsible for the out-of-plan instability of the ground state, the energy gap to them, the symmetrized atomic displacements along the coordinate of instability, and the energy profiles of these electronic states along this coordinate, as well as the vibronic coupling parameters characterizing the strength of instability.

Below in this section we demonstrate the methodology in the procedure of evaluating these parameters using specific molecular systems as examples. First we employ a series of hexa-heterocycles, (Figures 2 and 3), where the problem can be reduced to a two-level one [10]. Starting with the planar configuration, *ab initio* calculations of the electronic structure and vibrational frequencies of the system were carried out. All the four compounds, 1,4-$C_4N_2H_4X_2$ (X = H, F, Cl, Br) in Figure 3 have a nondegenerate ground electronic states A_{1g} and an imaginary frequency of b_{3u} symmetry with the values of 591.0, 554.5, 381.0, and 392.2 (in cm^{-1}) for X= H, F, Cl, Br, respectively. This means that the planar configuration of these compounds is unstable with regard to the symmetrized b_{3u} displacements of its atoms (the calculations [10] give also the directions and relative values of these displacements). According to the PJTE theory, the excited states that produce this instability of the ground state should have the B_{3u} symmetry. If we can assume that the main contribution to the instability come from only one low-lying excited state of B_{3u} symmetry, then the PJTE problem is simplified to a two-level one, $(A_{1g} + B_{3u})\otimes b_{3u}$. It is illustrated in Figure 3.

The next step in the evaluation of the PJTE is to reveal the energy gap to the active excited state and orbital excitations of the ground planar configuration that result in this excited state, as well as the parameters of the vibronic coupling between the two mixing states. For this reason ab initio calculations of the energy profiles (cross sections of the APES) of the systems along the instability coordinate b_{3u} were performed. The results are shown in Figure 4. We see that in all the four compounds the unstable along b_{3u} ground state $^1A_{1g}$ is accompanied by a relatively low-lying excited $^1B_{3u}$ state forming a PJTE interacting pair with curvatures assumed to follow the 2x2 secular equation (9) and the solutions (11). However, there may be higher-in-energy $^1B_{3u}$ states that contribute significantly to the instability of the ground state and, although less probable, it is not excluded that a higher excited is the main reason of instability (sought for in the process of manipulation). To estimate the possible contribution of higher excited states, a negative correction K_0' (to the primary force constants K_0) introduced by the latter is taken into account in the form of the second order perturbation [11] (see Eq. (6)):

$$K'_{01} = -2\sum_n \frac{|F_{1n}|^2}{E_n - E_1} Q^2 = -p_1 Q^2$$
$$K'_{02} = -2\sum_n \frac{|F_{2n}|^2}{E_n - E_2} Q^2 = -p_2 Q^2 \tag{15}$$

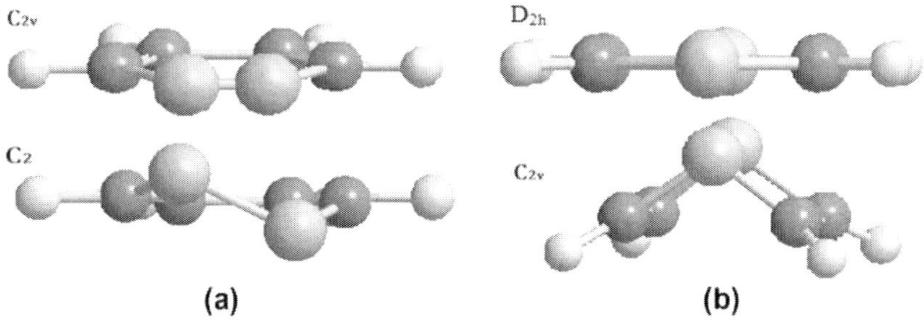

Figure 2. Side view of (a) 1,2-dichalcogenin planar (C_{2v}) and equilibrium twisted (C_2) structures, and (b) 1,4-dihalcogenin planar (D_{2h}) and equilibrium bent (C_{2v}) geometries.

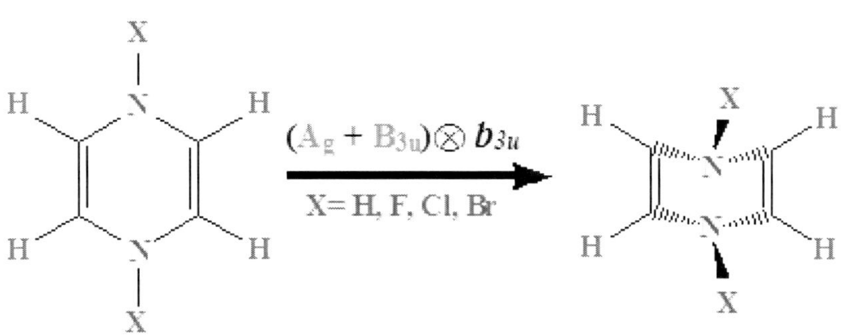

Figure 3. Schematic presentation of the b_{3u} type out-of-plan distortion produced by the $(A_{1g} + B_{3u})\otimes b_{3u}$ problem of the PJTE in the series of four hexa-heterocycles 1,4-$C_4N_2H_4X_2$ (X = H, F, Cl, Br).

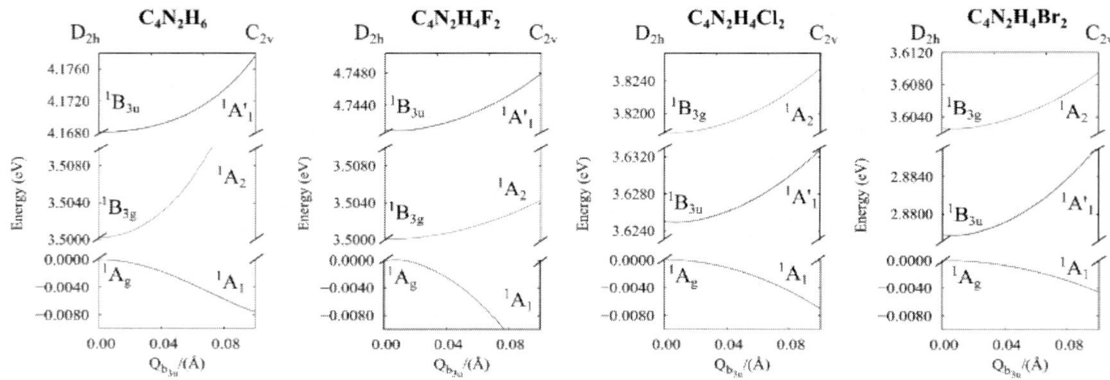

Figure 4. Energy profiles (cross sections of the APES) along the coordinate of instability b_{3u} of the planar D_{2h} configuration of the four hexa-heterocycle systems toward the distorted (buckled) C_{2v} structure.

We thus separate the whole spectrum of excited states contributions in two parts: the main contribution of the lowest excited states taken into account in a pseudo degenerate problem with a secular equation of the type (9), and the corrections from all the other excited states that lower the primary force constants of the two interacting states. This leads to the following secular equation of the two-level PJTE problem under consideration:

$$\begin{vmatrix} \frac{1}{2}(K_{01} - p_1)Q^2 - \varepsilon & FQ \\ FQ & \frac{1}{2}(K_{02} - p_2)Q^2 + \Delta - \varepsilon \end{vmatrix} \tag{16}$$

with the solution,

$$\varepsilon_{1,2} = \frac{1}{4}(K_{01} - p_1 + K_{02} - p_2)Q^2 + \frac{\Delta}{2} \pm \frac{1}{2}\sqrt{\left[\frac{1}{2}(K_{01} - p_1 - K_{02} + p_2)Q^2 - \Delta\right]^2 + 4F^2Q^2} \tag{17}$$

By fitting these equations to the ab initio calculated energy profiles in Figure 4 we get the parameter values listed in Table 1. It shows in detail (with numerical estimates) how the excited states distort the ground state planar configuration via the PJTE: all the four hexa-heterocycles under consideration become unstable (acquire negative force constants K) along the b_{3u} puckering distortion, shown in Figure 3. Although the separate calculation of the p_1 and p_2 corrections by the infinite sum in Eq. (6) is not possible, they can be extracted from the data in Table 1 by estimating the K_0 values after Eq. (3) using a special computer program [11]. But for the problem under consideration it is just important that K_{01}-p_1>0, showing that the contribution to the instability of the higher excited states (estimated by the negative value –p), although lowering the force constant, by themselves do not produce the instability of the ground state. This means that the lowest excited state $^1B_{3u}$ is responsible for the main (decisive) contribution to the instability, and hence the higher states are not the main target for manipulation. For simplicity, in many cases these *p* values are not evaluated numerically.

The calculations [10] show also the molecular orbital (MO) populations in the active ground and excited states to be targeted in attempts to suppress the PJTE in these compounds and to flatten their configuration. Actually, the MO presentation of the problem instead of (or in addition to) energy terms is important as a simplified, but chemically attractive (illustrative) description, which provides information about the role of individual orbitals in the origin of instability, and in many cases allows for qualitative estimate of the PJTE with much less computations. For the MO presentation, the electron-nuclear interaction part V in the Hamiltonian H in Eq. (7) is presented as a sum of one-electron functions $V = \Sigma_i V_i$, and the two wavefunctions ψ_0 and ψ_n in the definition of the vibronic coupling constant $F_{0n} = <\psi_0|(\partial H/\partial Q)_0|\psi_n>$ are taken in the form of single Slater determinants, composed by the one-electron wavefunction $|\varphi\rangle$ (MOs). Then by assuming that ψ_0 and ψ_n are the ground and excited states produced by an one-electron excitation $\alpha \rightarrow m$ from the occupied MO, $|\varphi_\alpha\rangle$, to the unoccupied MO, $|\varphi_m\rangle$, we get for the vibronic coupling constant F_{0n} in Eq. (7) the following expression:

$$F_{0n} = \langle \Psi_0 | (\partial H/\partial Q)_0 | \Psi_n \rangle = \sqrt{2} f^{\alpha m} \tag{18}$$

for singlet states, and $F_{0n} = \langle \Psi_0 | (\partial H/\partial Q)_0 | \Psi_n \rangle = f^{\alpha m}$ for doublet states, where

$$f^{\alpha m} = \langle \varphi_\alpha | (\partial V/\partial Q)_0 | \varphi_m \rangle \tag{19}$$

is the orbital vibronic coupling constants [5]. Similar to $F_{0n}, f_Q^{\alpha m}$ is nonzero only if the product of symmetries of the two MOs $|\varphi_\alpha\rangle$ and $|\varphi_m\rangle$ contains the symmetry of distortion Q.

According to Eq. (12), the contribution of the PJTE instability due to the vibronic coupling between these two orbitals is:

Table 1. PJTE coupling constants F_{12} (in eV/Å), energy gaps Δ, force constants with second order perturbation corrections K_1- p_1, K_2- p_2 (in eV/Å^2), and the total ground state force constant K of four heterocycle molecules in the planar configuration with respect to b_{1u} puckering distortions [10]

Molecules	K_{01}-p_1	K_{02}-p_2	F_{12}	Δ	K= (K_{01}- p_1)-(F_{12})^2/Δ
$C_4N_2H_6$	0.25	0.59	1.42	4.16	-0.23
$C_4N_2H_4F_2$	0.41	0.38	1.72	4.74	-0.21
$C_4N_2H_4Cl_2$	0.31	0.56	1.64	3.81	-0.40
$C_4N_2H_6Br_2$	0.29	0.63	1.56	3.60	-0.39

$$K_v^{0n} = -4\frac{(f^{\alpha m})^2}{E_n - E_0}, \tag{20}$$

if $|\varphi_\alpha\rangle$ is the double occupied MO (singlet state), and

$$K_v^{0n} = -2\frac{(f^{\alpha m})^2}{E_n - E_0}, \tag{21}$$

if $|\varphi_\alpha\rangle$ is a single occupied MO (doublet state). In general,

$$K_v^{0n} = -(q_\alpha - q_m)\frac{(F_{0n})^2}{E_n - E_0}, \tag{22}$$

where q_i are the occupation numbers of the mixing MOs.

Based on this orbital MO scheme, some qualitative conclusions can be made about the changes in the PJTE induced instability produced by external perturbations like oxidation, reduction, and excitation. In many heterocyclic systems the most significant negative contribution to the curvature of the ground-state APES in the direction of the active coordinate Q comes from the PJTE interaction with an excited state formed by the one-electron excitation from the highest occupied MO (HOMO) to an unoccupied MO of appropriate symmetry.

Then the removal of an electron from that HOMO reduces approximately by half the value of the negative vibronic contribution K_v in the oxidized state compared with the neutral molecule (see Eqs. (20) - (22)), thus suppressing the PJTE. Similarly, if the active excited state is formed by the one-electron excitation from one of the occupied MO to the lowest unoccupied MO (LUMO), then addition of an electron to this orbital by reduction decreases the value of K_v in reduced species, which leads to suppression of the PJTE, too.

The MO scheme can also be used to draw conclusions about suppression of the PJTE by excitations of the system. For instance, the one-electron excitation by removal of an electron from the HOMO followed by a single population of the LUMO may lead to the suppression of the PJTE if the LUMO is involved in the formation of the PJT active excited state, quite similar to oxidation or reduction (see Eq. (21). Note that the two PJTE active (ground-excited) MOs are coupled by symmetry; they are *not* directly related to bonding-antibonding MO pairs.

In a similar way, the suppression of the PJTE by the coordination of the system to any other atoms, molecules, or solids can be interpreted based on the change of orbital occupancy in the active MO pairs α-m due to the electron density transfer to the unoccupied MO m, or from the occupied orbital α. Such orbital charge transfers are considered in more detail in JTE and PJTE in systems with fractional charges [12]. With orbital charge transfers by coordination, the occupancies q_α and q_m in Eq. (22) become fractional; they are relatively easily obtained from electronic structure calculations. Applications of this approach to specific coordination systems has demonstrated its efficiency in rationalization of the experimental data on structural changes in molecular systems due to their interaction with another system [12, 13].

4. Methods of Suppression of the PJTE Induced Distortions in Heterocycles

4.1. Suppression of the PJTE in Heterocycles by Redox Influence and Excitation

4.1.1. Planarization of 1,4-Dithiine Containing Compounds

For examples to demonstrate the mechanism of suppression of the out-of-plane distortions in heterocycles induced by PJTE we consider first the thianthrene (dibenzo-1,4-dithiin) molecule, where the problem can be reduced to a two-level one [14]. From the crystal structure analysis it is known that the thianthrene molecule is bent relative to the S-S axis by 127°-128°, and it has low-energy flapping vibrations around this axis [15, 16]. On the other hand, the oxidized thianthrene molecule has a planar nuclear configuration with D_{2h} symmetry [17, 18]. Besides, if the carbon atoms in the 1, 4, 6, and 9 positions are substituted by nitrogen atoms, the dithiin core becomes planar [19].

Starting with the planar high-symmetry D_{2h} nuclear configuration, *ab initio* calculations of the geometry, electronic structure and vibrational frequencies of the system were carried out. In this configuration the molecule has a nondegenerate ground electronic state ^1A_g and an imaginary frequency of b_{1u} symmetry with the value 70.11 cm^{-1} (the symmetry axis of second order is perpendicular to the plane of the molecule). This means that the planar D_{2h} nuclear configuration is unstable with respect to the symmetrized *butterfly-like* distortion of b_{1u} type (directions and relative values of these displacements are shown in Figure 5). Therefore, according to the PJTE theory, the excited states that produce this instability should have the B_{1u} symmetry.

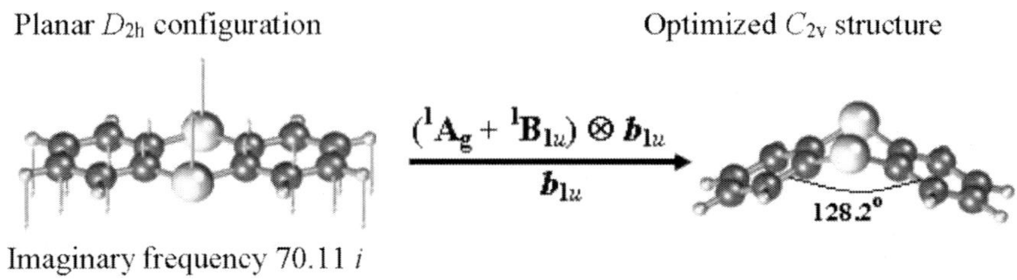

Figure 5. Puckering b_{1u} normal mode of the thianthrene molecule in the planar D_{2h} configuration, and the distorted equilibrium configuration produced by the PJTE problem $(^1A_g + {}^1B_{1u}) \otimes b_{1u}$. The arrow lengths show expected relative (not on scale) displacements.

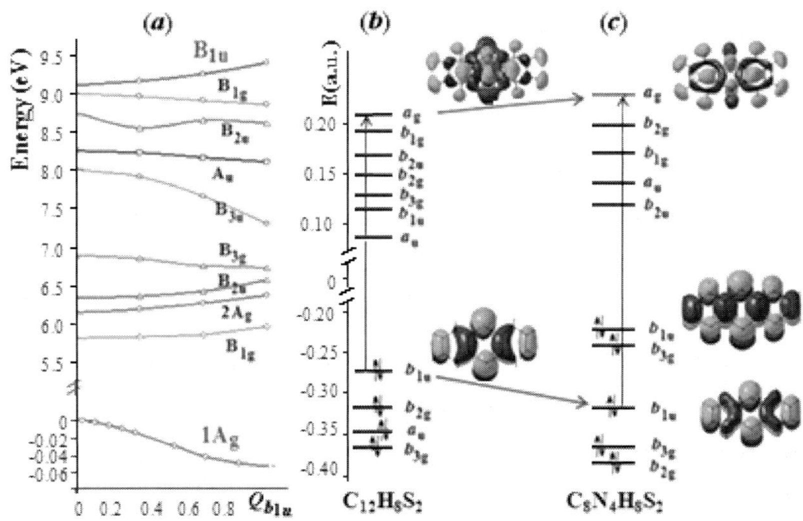

Figure 6. (*a*) Energy profiles (APES cross sections) along the b_{1u} distortion mode for the ground and excited states of the thianthrene molecule $C_{12}H_8 S_2$; (*b*) and (*c*), the MO energy level schemes for, respectively, the $C_{12}S_2H_8$ and $C_8N_4 S_2H_8$ molecules in the ground A_g state with indication of the one-electron excitations to the B_{1u} terms [14].

Further calculations [14] indicate that among the nine low-lying excited electronic states in the 10 eV energy domain, only one (at 9.17 eV) has the appropriate B_{1u} symmetry, active in the PJTE. It is formed by the one-electron excitation $b_{1u} \rightarrow a_g$ from the HOMO of b_{1u}-symmetry to the unoccupied MO of a_g-symmetry (Figure 6a). Hence the PJTE can be formulated as a two-level $(^1A_g + {}^1B_{1u}) \otimes b_{1u}$ problem. In order to evaluate the PJTE parameters, *ab initio* calculations of the energy profiles (cross sections of the APES) of the system along the instability coordinate b_{1u} were carried out. The results are shown in Figure 6a. By fitting the solutions of the secular equation of the type (17) to the *ab initio* calculated energy profiles, the following estimates of the PJTE parameters were obtained: K_{01}= 1.12 eV/Å^2, K_{02} = 0.96 eV/Å^2, F = 2.47 eV/Å, Δ_{01} = 9.17 eV. Near $Q \approx 0$, the ground state curvature becomes negative, $K \approx K_{01}$-$2F_{01}^2/\Delta_{01}$ = - 0.21 eV/Å^2, meaning instability of the planar configuration of the thianthrene molecule in its ground electronic state with respect to b_{1u} distortion, leading to the *butterfly-like* equilibrium geometry.

Then the removal of an electron from the HOMO by oxidation reduces the value of the vibronic constant F from $F = \sqrt{2}f^{a_g b_{1u}}$ to $F = f^{a_g b_{1u}}$, thereby decreasing the PJTE negative contribution K_v to the curvature of the APES ($K_v \approx -4f^2/\Delta$ for the neutral molecule compared with $K_v \approx -2f^2/\Delta$ for the radical cation). This results in the positive value for the curvature K of the APES, K= +0.58 eV/Å^2, indicating that the thianthrene cation should have a planar nuclear configurations. Indeed, calculations for the oxidized molecule [14] show that there are no imaginary frequency in the ground state of its planar configuration, that is, the planar geometry of the thianthrene cation is stable. This outcome is confirmed by experimental data [17, 18]. Moreover, calculations of the thiantrene molecule in its first excited state show that the excited molecule has a planar nuclear configuration (no imaginary frequencies). In this case, the excitation leads to the removal of one electron from the active (in the PJTE) HOMO b_{1u}, resulting in a decrease in the value of the vibronic coupling constant F.

As stated above, the suppression of the PJTE and restoration of the planar configuration takes place also by substitution of the carbon atoms at the 1, 4, 6, and 9 positions by the more electronegative nitrogen atoms. From comparison of the MO energy level schemes of the two systems, $C_{12}S_2H_8$ and $C_8N_4S_2H_8$ (Figs. 6(b) and 6(c)), it is seen that the energy gap between the mixing b_{1u} and a_g MOs, which produce the PJTE instability, increases significantly when moving from the former to the latter, reducing the PJTE. In addition, the emergence of two new upper occupied MOs (b_{3g} and b_{1u} in Figs. 6) increases the value of K_0. Calculations of this system in the D_{2h} nuclear configuration do not show any imaginary frequency, confirming that the planar configuration of the system is stable.

Thus, we conclude that 1) the folding of the neutral thianthrene molecule is due to the PJT coupling of the ground A_g state with the excited B_{1u} state, and 2) restoration of the planar nuclear configuration of this system by oxidation and excitation is directly related to the decrease in the vibronic coupling between these states, while substitution of C1, C4, C6, and C9 by N results in an increase in the energy gap between the mixing states. All of these external influences lead to the suppression of the PJT effect, resulting in the restoration of the planar configuration of the thianthrene molecule.

Following Refs. [10, 14, 20], the structural properties of other 1,4-dithiine containing heterocycles are similar to those considered above for thianthrene. All of them are unstable in the high-symmetry planar geometry, while in the equilibrium *butterfly-like* nuclear configuration they are bent at the S-S (N-N) axis. It was shown that these distortions are fully controlled by the PJTE, its mechanism being demonstrated in detail by ab initio calculation. In these systems the excited states, which provide for the main negative contribution to the curvature of the APES in the ground state, are mainly formed by the one-electron excitation from the HOMOs to one of the unoccupied MOs of appropriate symmetry. Therefore, as in the case of thianthrene, the cations of the above-mentioned compounds should have planar configurations. To verify this statement, calculations of the oxidized molecules were performed, and they show that indeed the planar geometry of the considered cation radicals is in equilibrium. The conclusion is that the restoration of planar nuclear configurations of the 1,4-dithiine containing molecules by oxidation is due to the decrease in the value of the vibronic coupling constants, which leads to the suppression of the PJTE and, consequently, to planarization of the molecules.

The redox influence on the PJTE instability and distortions was also explored in two tricyclic 1,2-dithiine containing systems, dibenzo[1,2]dithiine, $C_{12}H_8S_2$, and carbon sulfide C_6S_8, containing two S-atoms in the 1,2-positions of the central six-membered ring and one thione (C=S) bond in the five-membered rings on its either side. These compounds were studied both theoretically and experimentally [21, 22]. It was shown that the neutral molecules are not planar, they have a puckered structure of C_2 symmetry with a twisted 1,2-dithiin core, whereas a two-electron reduction resulted in the cleavage of the S-S bond and rotation of the two peripheral rings about the central C-C bond. It was shown also [9, 14] that these a_2-type out-of-plane distortions of all considered systems are triggered by the PJT coupling between the ground $^1A_1$ and two excited electronic states of $^1A_2$ symmetry, (resulting in the three-level $(^1A_1+1^1A_2+2^1A_2)\otimes a_2$ PJTE problem), while the reduction of the neutral molecules to $C_6S_8^{2-}$ dianions leads to enhancement of the PJTE and a much stronger distortion. Moreover, the oxidation of the $C_{12}H_8S_2$ molecule suppresses the PJTE and planarization of its cation $(C_{12}H_8S_2)+$, while in $C_6S_8^+$ the PJTE is weakened, being completely suppressed in $(C_6S_8)^{2+}$.

The possibility of flattening of puckered molecular systems was also checked-out directly, by *ab initio* calculations, performed on a series of tetra-heterocyclic 1,2-diazetes [23] $C_2N_2E_4$, E = H, F, Cl, Br (see also PJTE studies of tetrafluorocyclobutadiene and tetrasilacyclobutadiene analogues [24]). It was revealed that the stabilization of their planar configurations is achieved only in the dication states. The reason for the weakening of the PJTE in these cases is the much larger energy gaps Δ between the (changed) ground $^1A_1$ and active excited $^1A_2$ states. The other parameters in the condition of instability (14) include the primary force constant K_0, which increases by removing electrons (inter-electron repulsions *ceteris paribus* soften the system), thus weakening the PJTE too, while the vibronic coupling constant F formed by similar electronic states is expected to remain of the same order of magnitude.

4.1.2. Suppression of PJTE Induced Torsion Barriers to Rotation in Thiophene-Based Heterodimers

Another example of the suppression of the PJTE and planarization of heterocyclic compounds is presented by the reduction of torsion rotations in multi-ring thiophenes. These conformationally flexible molecules are studied extensively as possible units of polythiophenes used in organic electronics (see, e.g., [25] and references herein), where the problem of reducing the conformational flexibility and planarization of such systems is very topical. For this purpose two thiophene-based systems, bithiophene, $C_8H_6S_2$, and the model oligothiophene 2,5-bis(3-tetradecylthiophen-2-yl)thieno[3,2-b]-thiophene (BTTT) were considered [26].

The reference planar nuclear configuration of the *syn*-$C_8H_6S_2$ molecule (C_{2v} symmetry) is unstable with respect to the out-of-plane distortions of a_2 symmetry, corresponding to rotation of the two heterocycles around the central C_1-C_2 bond (Figure 7). In this configuration, the system has a nondegenerate ground electronic state $^1A_1$. Therefore, only the excited states of $^1A_2$ symmetry can produce the instability of the ground state. In Figure 8 the MO energy levels scheme and the calculated energy profiles along the instability coordinate a_2 for the ground $^1A_1$ and the low-lying excited $^1A_2$ states of $C_8H_6S_2$ molecule are presented. The first excited 1^1A_2 state is formed by the one-electron excitation from the HOMO of a_2 symmetry to the LUMO+1 of a_1 symmetry, the 2^1A_2 state corresponds to the excitation a_2(HOMO-1)→a_1(LUMO+1), while the 3^1A_2 state is formed by the one-electron excitation b_1(HOMO-3)→b_2(LUMO).

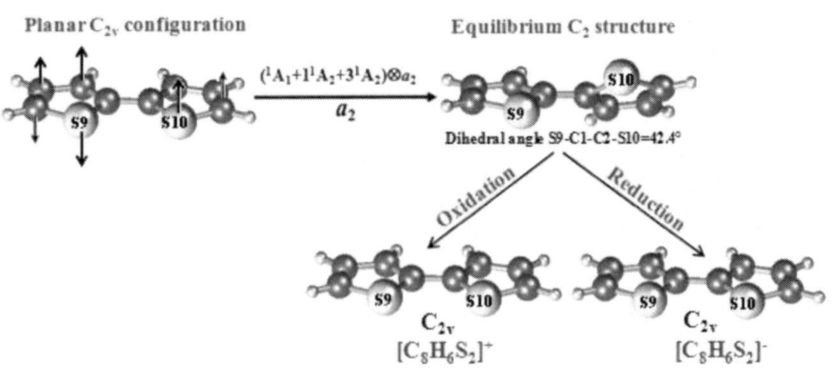

Figure 7. Twisting a_2 normal mode distortion of the planar C_{2v} configuration of the $C_8H_6S_2$ molecule induced by the PJTE problem (A_1+1A_2+3A_2)⊗a_2 (*syn* structure), and planarization of the equilibrium twisted structure by oxidation and reduction.

It was shown [26] that instability of the system with respect to torsion rotation is due to the three-level PJTE problem $(A_1+1A_2+3A_2)\otimes a_2$; the vibronic mixing of the second 2^1A_2 excited state with the ground state is shown to be less significant in comparison with the first and the third states, and vibronic contribution to the curvature from this excited state can be included in the primary force constant K_0. Fitting the secular equation to the calculated energy profiles yields the following values of PJTE parameters: $K_0 = 2.52$ eV/Å^2, $F_{01} = 2.41$ eV/Å, $F_{03} = 2.32$ eV/Å. Calculated energy gaps are: $\Delta_1 = 7.10$ eV and $\Delta_3 = 9.50$ eV. Then, near $Q \approx 0$, , $K \approx K_0 + K_{v1} + K_{v3}$, where $K_{v1} = -1.63$eV/Å^2, and $K_{v3} = -1.15$eV/Å^2. With these parameters the value of the curvature K of the APES is negative, K = -0.26 eV/Å^2.

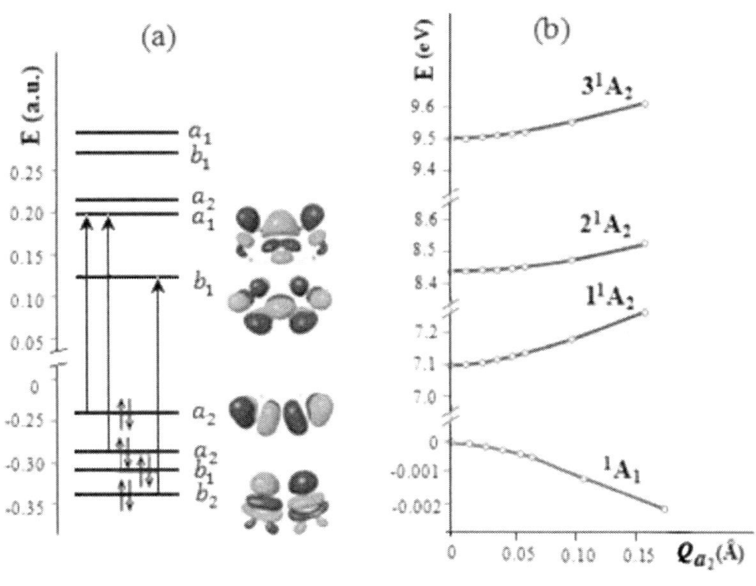

Figure 8. (a) MO energy levels scheme for the $C_8H_6S_2$ molecule in the ground A_1 state with indication of the one-electron excitations to A_2 terms; (b) *ab initio* calculated energy profiles along the a_2 twisting coordinate for the ground A_1 and three excited A_2 electronic terms. As the Q_{a2} coordinate, the values of the out-off-plan displacements of sulfur atoms were used [26].

Since the active 1^1A_2 and 3^1A_2 states are formed by including, respectively, the HOMO and the LUMO, the change in the populations of these boundary orbitals *via* oxidation or reduction suppresses the PJTE and may lead to flattening of the molecule. For instance, removing one electron from the HOMO decreases, approximately, by half the negative contribution to the curvature K_{v1} produced by the first $^1A_2$ excited state (see Eq. (22)), thus increasing the value of K. Similarly, an addition of one electron to the LUMO reduces the PJTE negative contribution K_{v3} provided by the third $^1A_2$ excited state that, too, increases the K value. With the obtained parameter values, estimation of the curvature of the APES in the cation and the anion species of the molecule yields: $K_{cat} \approx +0.54 eV/Å^2$, and $K_{an} \approx +0.32 eV/Å^2$, indicating that corresponding radicals should be planar [26]. Indeed, the analysis of the vibrational frequencies in the planar C_{2v} configuration of the cation $(C_8H_6S_2)^+$ and anion $(C_8H_6S_2)^-$ shows the absence of any imaginary frequency in both systems, i.e., the planar geometry of these charged species is the equilibrium one, as illustrated in Figure 7.

4.2. PJTE Suppression by Chemical Substitutions

4.2.1. Puckering and Planarization of 1,4-Dithiine S-Oxygenated Derivatives

The above example of planarization of the thianthrene molecule by atomic substitutions shows that chemical substitution of any atoms in heterocyclic compounds may lead to suppression of the PJTE that eliminates or reduces the out-of-plane distortions. On the other hand, substitution can also lead to an enhancement of the PJTE and subsequent symmetry breaking in the system. For example, the anthracene molecule (An) is planar with the nuclear configuration of D_{2h} symmetry, whereas the 9,10-dihydroanthracene (9,10-H_2An) has a folded geometry of C_{2v} symmetry with a dihedral angle between the planes of about 150-160° [27],. In the ground electronic state the anthracene molecule has no low-lying excited states of B_{1u} symmetry, which could induce the PJTE instability with respect to the folding of the molecule. But the addition of four hydrogen atoms in 9,10-H_2An produces an additional excited B_{1u} state, its vibronic contribution to the ground state resulting in the enhancement of the PJTE and the distortion [14].

The suppression of the PJTE by substitution was shown to take place in the S-oxygenated derivative of 1,4-dithiine, $C_4H_4(SO_2)_2$, which is planar [20]. There is no *butterfly-type* puckered structure in this molecule because the significant increase of the energy gap between the ground and the PJT active excited states, while the vibronic coupling constant is small, so the condition of instability is not satisfied [20]. On the other hand, DFT calculations [28] have shown that the derivatives of 1,4-dithiine with one and two SO groups are puckered and have a shadow boat conformations. It was shown [9] that the puckering and planarization of S-oxygenated derivatives of the1,4-dithiine molecule are due to, respectively, enhancement and suppression of the PJTE.

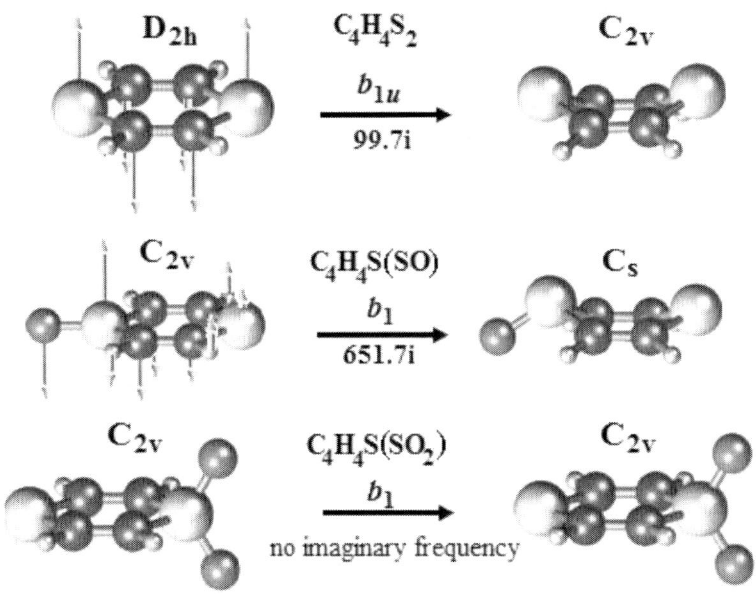

Figure 9. Puckering b_{1u}/b_1 normal modes in 1,4-dithiine and in its two oxygenated derivatives $C_4H_4S(SO)$, and $C_4H_4S(SO_2)$; the arrow lengths show expected relative (not on scale) displacements.

To demonstrate how changes in the electronic structure in a series of derivatives influence the PJTE, two cases of substitution, S→SO and S →SO_2, are discussed below (see Ref. [9]). Starting with the planar-ring high-symmetry nuclear configurations, *ab initio* calculations of the electronic structure and vibrational frequencies of the systems were carried out. As shown in Figure 9, the derivative with the SO_2 group maintains the planar-ring conformation (no imaginary frequencies), while the S→SO substitution leads to the symmetry breaking and ring puckering along the distortion coordinate of b_1 symmetry.

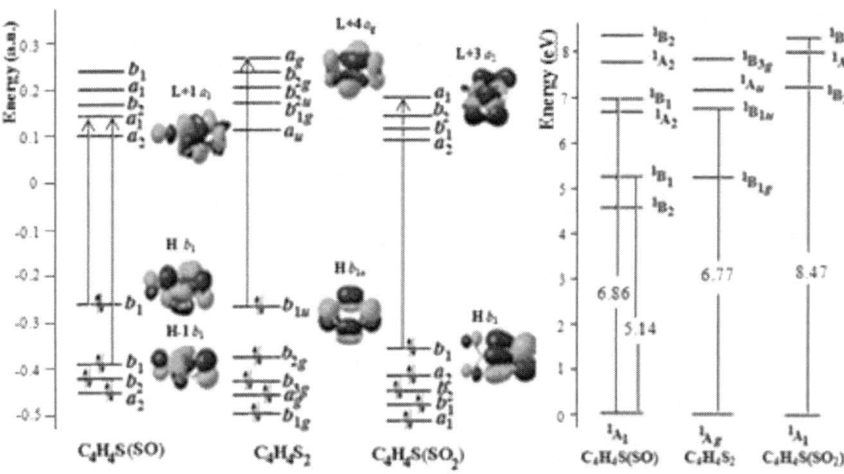

Figure 10. The left three schemes: MO energy levels scheme for the $C_4H_4S_2$, $C_4H_4S(SO)$, and $C_4H_4S(SO_2)$ molecules in their ground states in the planar-ring nuclear configurations, with indication of the one-electron excitations to excited B_{1u}/B_1 terms; the right three schemes: *ab initio* calculated energy gaps between the ground and excited states.

The origin of these structural features can be elucidated by a comparative analysis of the molecular orbital schemes and the relative energy gaps to the excited states of the molecules under consideration (Figure 10).

One can see that the substitution S→SO in the $C_4H_4S(SO)$ molecule donates electron density to the ring, which leads to the formation of one more excited state of B_1 symmetry, which, in turn, enhances the PJTE. On the contrary, the substitution $S \rightarrow SO_2$ leads to two effects, each of which suppresses the PJTE. Firstly, the highest occupied MO b_1 does not contain any contribution from the π-atomic orbital of the sulfur atom, which reduces the value of the vibronic coupling constants. The second effect is due to a significant increase in the energy gap between the ground and the excited state. Both these factors, increase of the energy gap and decrease of the vibronic coupling constants, lead to the suppression of the PJTE in the $C_4H_4S(SO_2)$ molecule, and to restoration of the planar-ring configuration.

We can conclude that the substitution accompanied by electron donation into the ring enhances the PJTE, while the substitution accompanied by the outflow of the electron density from the puckered ring leads to suppression of the PJTE and planarization of the ring (Figure 11).

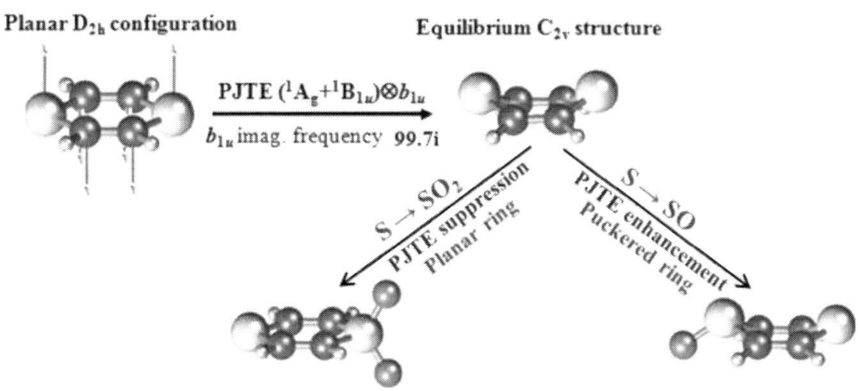

Figure 11. The PJTE in S-oxygenated derivatives of 1,4-dithiine: the planar configuration is puckered due to the $(^1A_g + ^1B_{1u}) \otimes b_{1u}$ PJTE, which is suppressed by $S \rightarrow SO_2$ substitution, but enhanced by the S→SO substitution.

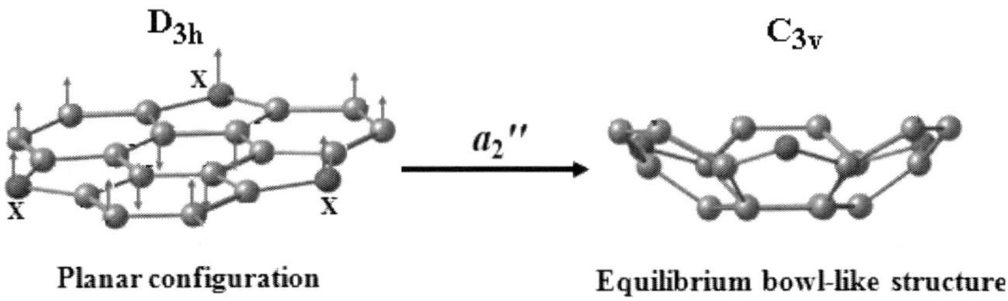

Figure 12. Puckering normal mode a_2'' in heterosumanenes in the planar D_{3h} nuclear configurations.

4.2.2. Substitutions in Heterosumanenes

Sumanene is a typical representative of the class of compounds known as molecular bowls, or buckybowls [29, 30]. The replacement of skeletal carbon atoms in heterosumanenes by isoelectronic substituents was shown to be most effective in modulating the bowl-to-bowl inversion barrier, and the consequent physical-chemical properties, such as the ability to adsorb small molecules [31, 32]. The analysis of the PJTE in these systems [26] shows how tri-atomic substitutions in sumanenes influence their buckling instability, especially the intriguing origin of the difference between tri-oxygen and tri-sulfur substitutions, the former being much stronger than the latter. These data are important also in view of possible applications that may require manipulation of the bowl structure by means of external (targeted) influence of the PJTE parameters.

The calculation of three $C_{18}X_3H_6$ (X=CH_2, O, S) molecules in the planar nuclear configuration of D_{3h} symmetry showed that in this configuration there is only one imaginary frequency corresponding to the out-of-plane distortions of a_2'' symmetry, which transforms the planar D_{3h} nuclear configurations into the equilibrium buckybowl structures of C_{3v} symmetry [26]. The puckering normal mode a_2'' is shown in Figure 12.

All the considered molecules in the planar nuclear configurations of D_{3h} symmetry have a non-degenerate ground electronic states, $^1A_1'$. Since the instability coordinate transforms as a_2'', only excited states of the A_2'' symmetry can contribute to the instability of the ground state, $F^{(0,i)} = \left\langle A_1' \middle| (\partial H/\partial Q_{a_2''})_0 \middle| A_2'' \right\rangle \neq 0$. These A_2'' excited state can be formed by one-electron excitations $e''_{xz} \rightarrow e'_x$ or $e'_x \rightarrow e''_{xz}$; in these cases the orbital vibronic coupling constant $f = \langle e''_{xz}|(\partial H/\partial Q_z)_0|e'_x\rangle \neq 0$ (see Eq. (19)). The cross sections of the APES for the sumanene molecule along the a_2'' distortion coordinate were calculated for the ground and four low-lying excited states of A_2'' symmetry (Figure 13). It was shown [26] that the most essential vibronic contribution to the curvature of the APES comes from the vibronic mixing between the ground and the third $3A_2''$ and fourth $4A_2''$ excited states; the vibronic contributions from the first and the second excited states are insignificant. The same result is obtained also for trioxosumenene and trithiosumanene. Hence, the three-level PJTE problem $(A_1' + 3A_2'' + 4A_2'')\otimes a_2''$ was explored for these cases. The fitting of the secular equation to the calculated energy profiles gives the parameters of the PJTE coupling, as presented in Table 2.

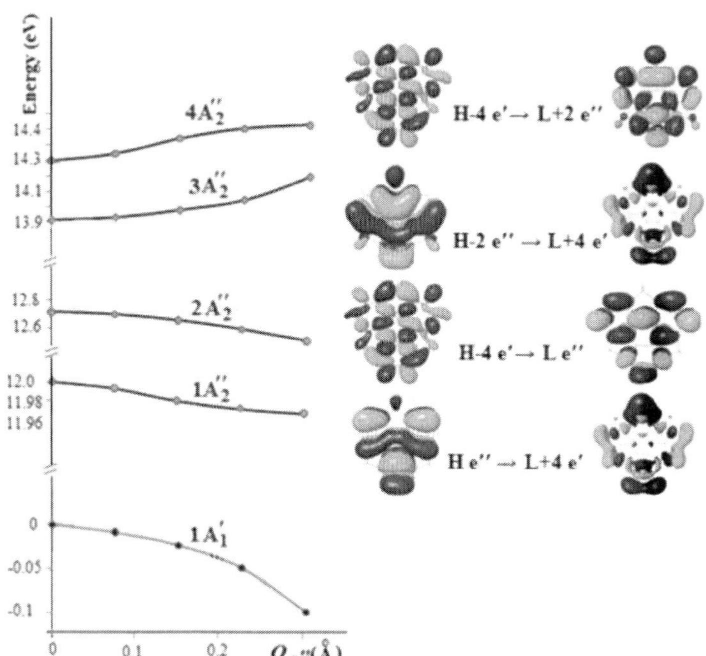

Figure 13. Ab initio calculated energy profiles for the ground A_1' and for four excited A_2'' electronic terms of the sumanene molecule along the a_2'' out-of-plane displacements. As the $Q_{a2''}$ coordinate, the bowl depth is used [26].

Table 2. Energy gaps Δ_{0i} (in eV), PJTE coupling constants F_{0i} (in eV/Å), primary force constants K_0 (in eV/Å^2), vibronic contributions to the curvature of the APES K_v (in eV/Å^2), and the full force constants K (in eV/Å^2) in the reference D_{3h} configuration of sumanene, trioxasumanene, and trithiasumanene molecules

Molecule	K_0	Δ_{03}	F_{03}	K_{v3}	Δ_{04}	F_{04}	K_{v4}	$K_v = K_{v3} + K_{v4}$	$K = K_0 + K_v$
Sumanene, $C_{21}H_{12}$	8.60	13.91	6.03	-5.23	14.32	6.51	-5.90	-11.13	-2.53
Trioxasumanene, $C_{18}O_3H_6$	7.8	12.1	7.21	-8.59	13.8	6.95	-7.00	-15.59	-7.79
Trithiasumanene, $C_{18}S_3H_6$	8.8	11.1	5.05	-4.60	12.2	5.24	-4.50	-9.10	-0.30

The main conclusions from these data are as follows [26]. The instability of the planar nuclear configuration of all the considered sumanenes is produced by the PJTE $(A_1' + 3A_2'' + 4A_2'')\otimes a_2''$. In the case of trioxosumenene the energy gaps between the ground and the excited A_2''states are smaller compared with the sumanene molecule, which enhances the PJTE. On the other hand, the values of the vibronic coupling constants increase, because the contributions of the oxygen 2p-AOs in the active molecular orbitals are larger than those of the benzylic carbon atoms in the sumanene molecule. Both these factors, decrease of the energy gaps and increase of the vibronic coupling constants, lead to enhancement of the PJTE in the trioxosumanene molecule, which results in a stronger distortion (Figure 14).

In the trithiasumanene molecule, substitution of the oxygen atoms by sulfur atoms leads to two opposite effects. Firstly, the energy gaps in this case are somewhat smaller compared to the trioxosumanene molecule. However, the main effect of this substitution is the significant decrease in the values of the vibronic coupling constants (see Table 2) that leads to the suppression of the PJTE. This is because the orbital vibronic coupling constant is proportional to the derivative of the overlap integral of atomic orbitals of near-neighboring atoms, which characterizes the rate of change of interatomic coupling under distortion [5, 6]. In the case under consideration, the changes in overlap of the more diffuse 3p-orbitals of the sulfur atoms with the 2p-AOs of near-neighbor carbon atoms are significantly slower than that between the 2p-AOs of the oxygen and carbon atoms.

To summarize, tri-atomic substitutions in sumanenes influence their puckering instability; more electronegative oxygen atoms enhances the PJTE due to the donation of additional electronic charge to the π-system of the sumanene molecule resulting in a stronger distortion, whereas in trithiasumanene $C_{18}S_3H_6$, the orbital vibronic coupling constants decrease and the PJTE is suppressed, resulting in a smaller distortion (Figure 14). The same effect of flatening of the frameworks is expected to take place in trisilasumanenes and trigarmasumanenes [33]. Spontaneous symmetry breaking in cyclo[18]carbon induced by the PJTE is analyzed in Ref. [34].

4.2.3. Symmetry Breaking in Halogenabenzenes

The PJTE origin of symmetry breaking in halogenabenzenes $(CH)_5X$ (X = F, Cl, Br, I) and in $C_6H_7^-$, as well as the possibility to restore their planar configurations was considered in Ref. [35]. It was shown that their planar configuration of C_{2v} symmetry is unstable with respect to the out-of-plane b_1 distortion leading to equilibrium bend structure of C_s symmetry (Figure 15).

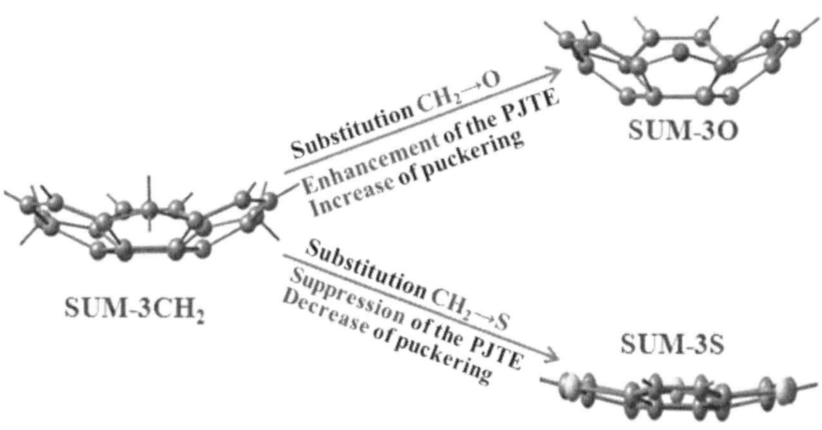

Figure 14. Enhancement (trioxosumanene) and suppression (trithiosumanene) of the PJTE by substitutions in heterosumanenes.

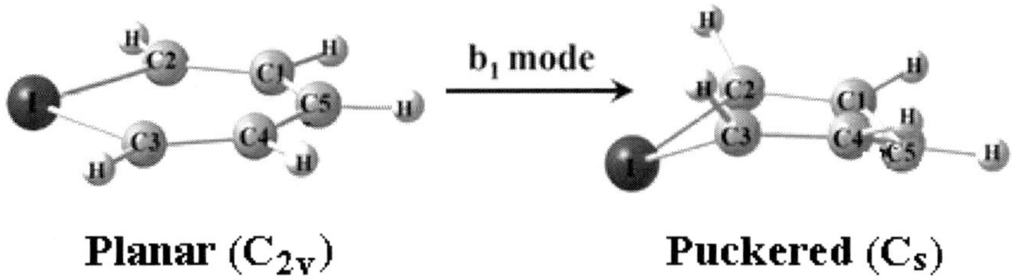

Figure 15. The planar (unstable) configuration and the PJTE induced distorted structure of $(CH)_5I$.

The reason for such distortion is the vibronic coupling of the ground $^1A_1$ state with two excited states of $^1B_1$ symmetry. The $(CH)_5F$ molecule is an exception, it has only one active $^1B_1$ excited state, but two imaginary frequencies corresponding to two b_1-type symmetry displacements in the planar configuratiomn. For this case, the two-mode PJTE problem has been solved resulting in the single interaction mode $Q_{b1} = 0.57\ Q_{1b1} + 0.82\ Q_{2b1}$. It is also stated [35] that the flattening of halogen-benzenes can be achieved by removing the two electrons from the HOMO of π^*-type or by substituting hydrogen atoms adjacent to halogen atoms with ultra-strong electron-withdrawing BH_3 groups or other isolobal transition metal fragments.

4.3. Flattening Puckered (Buckled) Cyclic Structures and Their Extended 2D Systems by Coordination

In comparison with the methods of flattening puckered heterocyclic systems by suppression of the PJTE via oxidation or substitution, considered above, coordination to external atoms or molecular groups with the same purpose seems to be less invasive and in many cases practically more attractive, especially in application to 2D systems. Theoretical considerations and possible quantitative estimates of the effect can be obtained based on presentation of the PJTE in terms of MOs and fractional charge transfers [12], shown in section 3 (Eq. (22)). Below we present several illustrative examples of flattening cyclic compounds and their extended 2D systems by suppressing the PJTE. More examples see in [8].

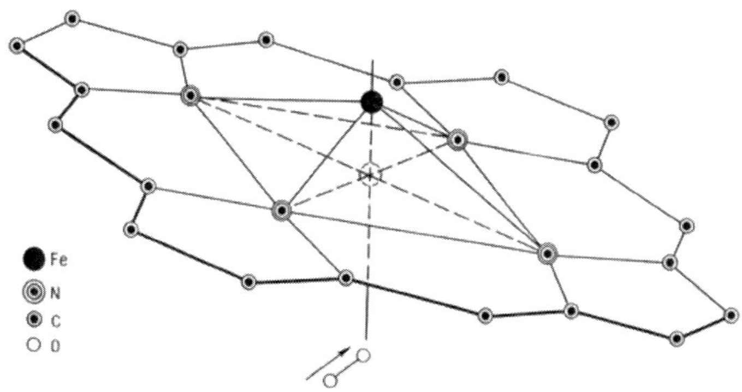

Figure 16. Iron porphyrin ring (in hemoglobin) with the iron atom out-of-plane displaced due to the PJTE. The planar configuration (shown by dashed line) is reached by coordination of the oxygen molecule (shown by arrow), which suppresses the PJTE [37].

4.3.1. Metal-Porphyrin Ring in Hemoglobin Oxygenation

First example of flattening a buckled molecular ring system by suppressing its PJTE-induced out-of-plane distortions was revealed as taking place in hemoglobin oxygenation [36, 37]. It was shown that in free deoxy-hemoglobin iron-porphyrin centers (Figure 16) the iron atom is out-of-plane displaced from its porphyrin ring due to the PJTE, and returns back to the in-plan position by oxygenation (triggering the T→R conformational transition), because the coordination of the oxygen molecule destroys the condition of the PJTE instability, Eq. (14). In this case the main PJTE parameter influenced by outside coordination is the energy gap Δ between the ground state (composed mainly by the nitrogen atomic *p* functions) and active exited electronic state (empty d_{z^2} orbital of iron): coordination of the oxygen molecule changes this energy gap from Δ ~ 0.2 eV to Δ ~ 3 eV, thus reversing the inequality (14) [36, 37]. Similar situations occur also in many other planar systems with transition metals participation (a similar effect may take place in 2D transition metal dichalcogenids).

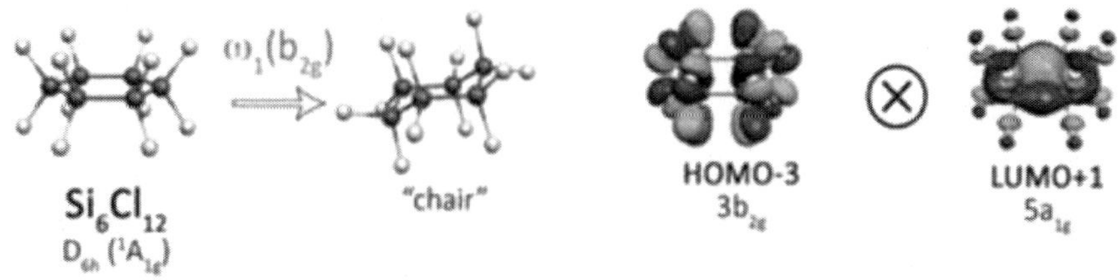

Figure 17. Illustration to the PJTE distortion of the high-symmetry D_{6h} configuration of the Si_6Cl_{12} compound in the ground $^1A_{1g}$ state showing the interacting pairs which produce the PJTE instability: the b_{2g} type normal displacements toward the "chair" configuration of the system is facilitated by the vibronic coupling between the (HOMO-3) $3b_{2g}$ and (LUMO+1) $5a_{1g}$.

4.3.2. Cyclohexasilane Systems Si_6X_{12} (X = Cl, Br).

In the study of cyclohexasilane systems [38] Si_6X_{12} (X = Cl, Br), it was shown that both these compounds are unstable in the high-symmetry D_{6h} geometry with a planar S_6 ring, and acquire a chair-like structure in the stable D_{3d} configuration (similar to C_6H_{12}), shown in Figure 17. Detailed electronic structure calculations of the Si_6X_{12} systems in the ground and several excited states in the D_{6h} configuration reveals their occupied and unoccupied molecular orbitals, the vibronic coupling between which makes this configuration unstable with respect to lower symmetry nuclear displacements. Figure 17 illustrates these findings for Si_6Cl_{12}.

In terms of molecular orbitals it shows the coupled (HOMO-3) and (LUMO+1) that realize the PJTE instability with respect to corresponding low-symmetry displacements. What is not seen directly from Figure 18 is that the indicated b_{2g}-$5a_{1g}$ MOs are orthogonal in the undistorted configuration, and become non-orthogonal under the indicated low-symmetry displacements of the atoms. Their nonzero overlap produces additional covalence bonding, which is the main driving force of the PJTE distortion. The latter occurs only when the contribution of additional covalency overcomes the stiffness of the system in the direction of the distortion represented by the primary force constant K_0 in the condition of instability (14). This explains the origin of the distorted configuration of the Si_6Cl_{12} molecule in equilibrium.

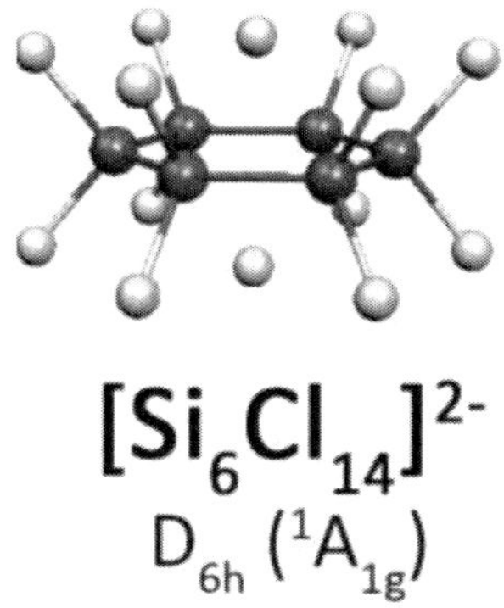

Figure 18. The structure of the dianion $[Si_6Cl_{14}]^{2-}$ in an inverse sandwich arrangement, in which the basic Si_6Cl_{12} fragment is in the high-symmetry D_{6h} configuration due to the suppressed PJTE by the coordinated two Cl^- ions; the free Si_6Cl_{12} system has a chair D_{3d} symmetry in equilibrium due to the PJTE distortion.

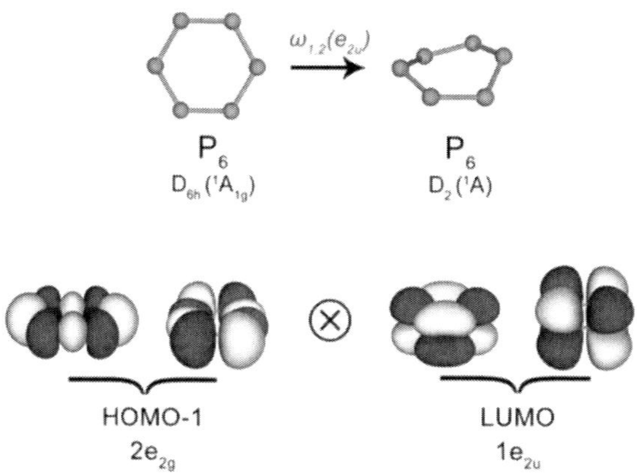

Figure 19. The PJT vibronic interaction of the HOMO e_{2g} with the LUMO e_{2u} of the D_{6h} ($^1A_{1g}$) configuration of P_6 is responsible for its e_{2u} distortion toward the D_2 (^1A) structure (Reprinted with permission from Ref. 42. Copyright 2012 American Chemical Society).

However, when a dianion $[Si_6Cl_{14}]^{2-}$ was isolated [39] it emerged that it has an inverse sandwich structure with a flat planar S_6 ring and two Cl^- ions below and above the mid-ring (Figure 18). It follows that the two donor ions flattened the puckered ring and restored its high-symmetry D_{6h} configuration. Ab initio calculations [38] reveal in detail the mechanism of this $D_{3d} \rightarrow D_{6h}$ transformation of the Si_6Cl_{12} system under the influence of two symmetrically coordinate Cl^- ions: the excess electrons of the latter occupy the active unoccupied state $5a_{1g}$ (Figure 17) making it the HOMO state, thus *suppressing the PJTE* that causes the instability of the D_{6h} configuration (other symmetry allowed excited LUMO states that could produce the PJTE are too high in energy). To further confirm this effect experimentally the authors [38] synthesized and analyzed a series of similar compounds, $[^nBu_4N]_2[Si_6Cl_{12}I_2]$, $[^nBu_4N]_2[Si_6Br_{12}I_2]$, $[^nBu_4N]_2[Si_6Br_{14}]$; in all of them the Si_6X_{12} fragment has the highest D_{6h} symmetry, as expected due to the suppressed PJTE. Ring flattering in $C_5H_5^-$ and $[Si_5Cl_{10}{\bullet}2Cl]^{2-}$ is discussed in Refs. [40, 41].

4.3.3. Triple-Decker Sandwich Structures CpMoE$_6$MoCp (E = P, As, Sb; Cp=C$_5$H$_5$).

A similar to the outlined above methodology is employed in the reported analysis of the triple-decker sandwich structures of a series of complexes [42] $CpMoE_6MoCp$ (E = P, As, Sb; $Cp=C_5H_5$). In these systems, the E_6 groups were shown to occupy the middle-deck position with a planar regular-hexagonal configuration. Meanwhile, these free E_6 groups may hardly be planar. The calculations show that they have several distorted configurations with much lower energy; for P_6 the lowest distorted configuration is by 34.9 kcal lower than the regular hexagonal one. As expected, the distortion of the latter is due to the PJTE. Figure 19 shows the interacting HOMO and LUMO that produce the e_{2u} distortion of the P_6 planar hexagonal ring of phosphorous.

It follows that in the triple-decker sandwich structure where the E_6 fragments become planar, the PJTE in the latter is fully suppressed, as illustrated in Figure 20. The detailed *ab initio* calculations with geometry optimization confirm this conclusion showing that the active LUMO responsible for the PJTE distortion of the E_6 ring (Figure 19) become occupied by the molybdenum electrons. [42].

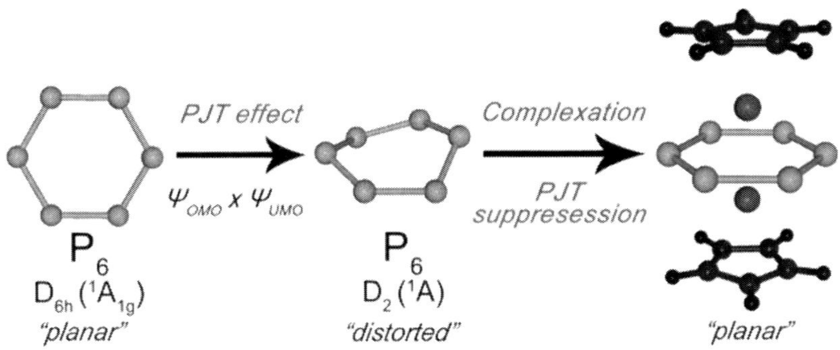

Figure 20. Illustration to the suppression of the PJTE in the P_6 ring by its complexation in a triple-dekker system $CpMoE_6MoCp$ (Reprinted with permission from Ref. 42. Copyright 2012 American Chemical Society).

4.3.4. Suppression of the PJTE by Aggregation in 1,4-Dithiines

Another mechanism of suppressing the PJTE in heterocyclic systems by aggregation due to intermolecular p-stacking interactions of planar configuration was proposed in [43]. Figure 21 illustrates this mechanism. Performing detailed ab initio calculations the authors show that the p-staking interactions between parallel planar configuration of the C_6S_8 molecule contributes to the weakening of the PJT distortions in each of them, making them less distorted, and reaching an exact planar configuration in a trimer parallel arrangement shown in Figure 21. This explains the origin of the $C_{2h} \rightarrow C_2$ transformation of the molecular geometry when moving from the free molecule to the α polymorph of the crystal state. The C_6S_8 compound in the solid state has also another polymorph, β, for which the distorted configuration of the free molecule remains distorted in the crystal structure.

A similar phenomenon of flattening by aggregation takes place in a more extended quasi planar compound dithianon, $C_{14}H_4N_2O_2$-S_2, which is distorted in the free molecule, but becomes flattened by aggregation in one of its polymorphs. In this case, as shown by calculations [43], the parallel stacking of only two molecules is sufficient to destroy the PJTE induced out-off-plan distortion.

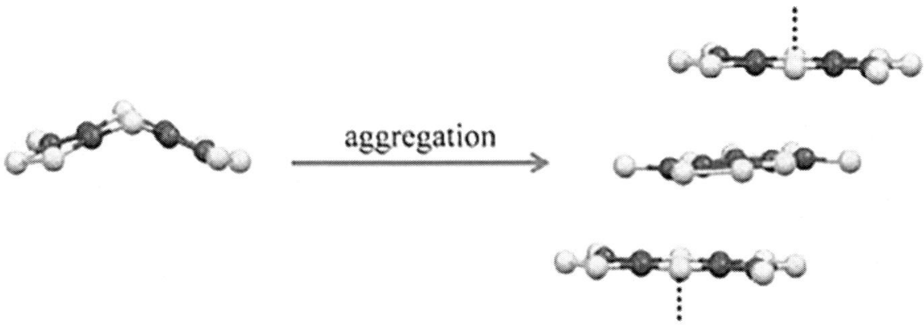

Figure 21. The free molecule of 1,4-dithiine containing compound, C_6S_8, (left) has a puckered structure of C_2 symmetry (induced by the PJTE in the planar high-symmetry configuration C_{2h}), which is fully suppressed in the crystal structure of the a polymorph due to the p-stacking interaction between at least three molecules in a planar-parallel arrangement (right).

4.3.5. Buckling of Triazine-Based Graphitic Carbon Nitride (g-C_3N_4)

Graphitic carbon nitride g-C_3N_4 is a 2D system is similar to graphene in many aspects, but unlike graphene g-C_3N_4 is not regular planar, it is buckled as shown in Figure 22 for a whole sheet and in Figure 23 in a cluster unit presentation [44]. The PJTE origin of the buckling in the cluster unit $C_9N_{12}H_6$ (Figure 23) was studied in detail with the goal to influence it by external perturbations [44]. It was shown that its planar configuration is unstable along the e″ symmetrized displacements for which the coupled HOMO and LUMO, responsible for the instability, were revealed by ab initio calculations. With these data and previous experience, the authors [44] suggested suppressing the PJTE and restoring the planar configuration of this system by adding Be^{2+} ions below and above the fragment unit in an inverse sandwich arrangement, similar to the above cyclohexasilane case. The calculations show, indeed, that the electron acceptor ions change the HOMO-LUMO positions thus quenching the PJTE and flattening the system. Figure 23 shows the optimized structures of both the fragment unit and the extended sheet of g-C_3N_4 with appropriately positions of the Be^{2+} ions, which restore the planar configuration.

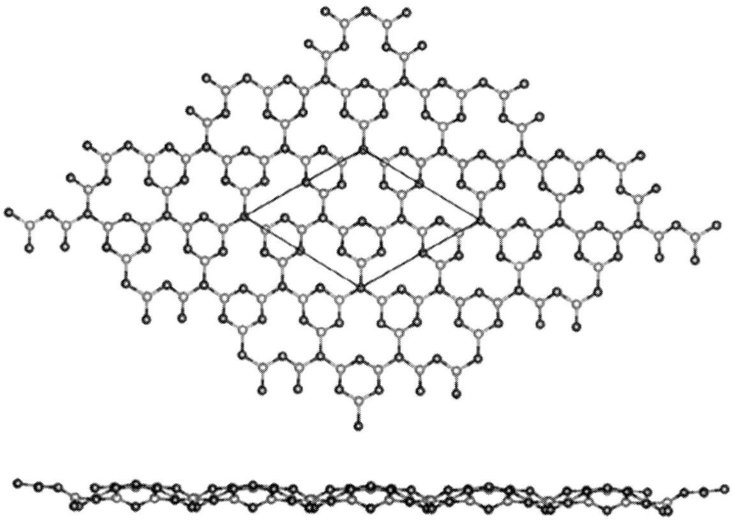

Figure 22. Buckled 2D structure of graphitic carbon nitride g-C_3N_4 shown in top and side view. Here and below the gray, blue, and white spheres represent carbon, nitrogen, and hydrogen atoms, respectively (Reprinted with permission from Ref. [44]. Copyright 2015 American Chemical Society).

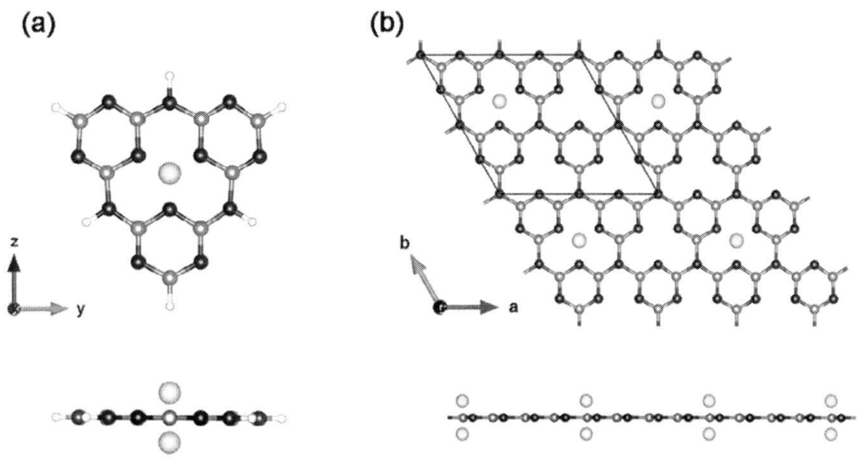

Figure 23. Top and side views of the optimized complex of the graphitic carbon nitride, g-C_3N_4, local unit with two Be^{2+} ions in an inverse sandwich arrangement (a), the corresponding extended 2D sheet (b) [44]. (Reprinted with permission from Ref. [42]. Copyright 2015 American Chemical Society).

4.3.6. 2D Silicene

Silicene is the nearest analog of graphene, and therefore it got elevated attention in search for useful 2D materials. Similar to graphene, silicene has a hexa-ring structure, but distinguished from graphene, the silicene rings are not planar, but puckered. As expected based on the general theory, the puckering of the planar configuration of silicene was shown to result from the PJTE, the latter being the basic reason of its structural difference from grapheme [45, 46].

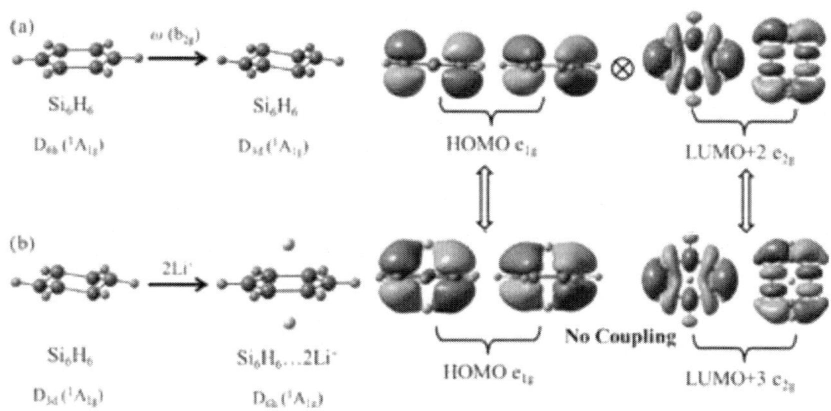

Figure 24. (a) Onset of the puckering distortion $D_{6h} \rightarrow D_{3d}$ via b_{2g} displacements in hexasilabenzene, and the HOMO and LUMO involved in the PJTE; (b) Suppression of the ring distortion in the hexasilabenzene-2Li$^+$ complex: with the HOMO and LUMO of same symmetry, the PJTE is suppressed (shown as "no coupling") because of the significant increase of the energy gap between them (Reprinted with permission from Ref. 46. Copyright 2014 American Chemical Society).

Figure 24 illustrates the results obtained for the analog of the silicene local structure, hexasilabenzene Si_6H_6, which has the same hexagonal Si_6 ring [46]. The PJT vibronic coupling between its HOMO e_{1g} and LUMO e_{2g} results in the b_{2g} type distortion transforming the D_{6h} structure into a D_{3d} one (Figure 24a). Then they show that two positive ions Li^+ arranged in an inverse sandwich geometry fully suppress the PJTE distortion (Figure 24b). The calculations [46] show also that the interaction of the Li^+ ions with the silicon ring is mostly ionic, and the PJTE is suppressed because of the increase of the HOMO-LUMO energy gap Δ between the PJT interacting states from 6.83 eV to 7.68 eV, which destroys the instability criterion Eq. (14). The authors [46] discuss also the implications that occur in the extended systems and suggest a virtual (an *ab initio* optimized) structure of silicene with intercalated Li^+ ions (Figure 25), in which the PJTE is suppressed and the Si atoms form regular hexagons in a planar arrangement, similar to graphene. Mixed carbon-silicon ring systems are also considered [46].

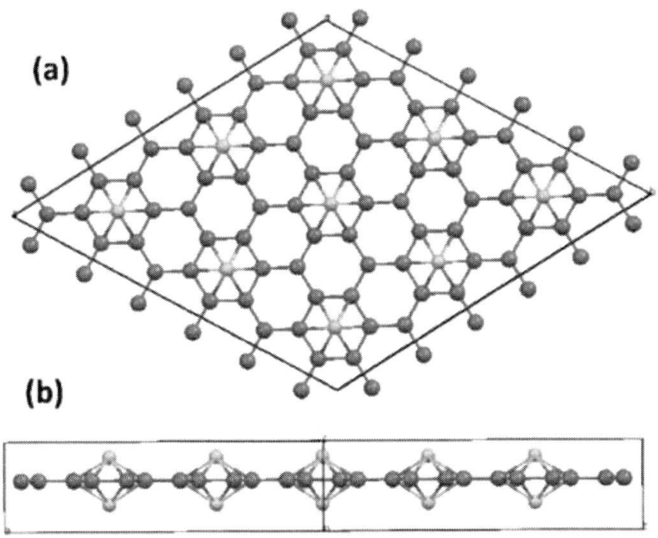

Figure 25. Optimized structure of silicene with appropriately intercalated Li^+ ions: (a) transverse view, (b) lateral view (Reformatted with permission from Ref. 46. Copyright 2014 American Chemical Society).

4.3.7. 2D Phosphorene

Phosphorene (Pn) can be viewed as a layered quasi 2D system composed by buckled hexagonal rings. The relatively heavy P atoms in Pn produce rather small energy gaps to excited states resulting in several stable conformations of lower symmetry [47, 48]. Among the latter the two stable buckled configuration known as black and blue phosphorene are most important in view of the 2D systems under consideration. The buckling of Pn can be conveniently studied using the P_6H_6 molecule as a model unit [47]. The planar regular-hexagonal D_{6h} configuration of P_6H_6 is unstable due to the PJTE; as shown by calculations, there are two most active PJT displacements, e_{1g} and b_{2g}, leading to two stable distorted configurations of C_{2h} and D_{3d} symmetry. It is notable that the basic P_6 skeletons in these two molecular configurations of P_6H_6 coincide well with the ones in black and blue phosphorene, respectively (Figure 26). Hence the PJTE origin of the buckling in the molecular system can be employed to explain the origin of the buckling in the extended systems of black and blue Pn and to manipulate their properties.

The authors [47] suggested using several molecular systems (shown in Figure 27) as molecular dopants in order to influence the structural properties of the two phosphorenes by targeting their PJTE parameters of the latter with the known donor-acceptor properties of the former (molecules 1–4 are known to be donors, while 5 and 6 are acceptors). To explore this possibility, calculations of charge distribution and charge transfers by coordination (adsorption) of these molecules to the surface of black and blue phosphorene, as well as their influence on the electronic band structure and band gaps, were carried out [47] (see also [49]).

It emerges from these data and electronic band structures that the dopant molecules indeed influence significantly the properties of the Pn systems by either donating (molecules 3 and 4) or accepting (molecules 5 and 6) electronic charge, thus modifying the PJT interacting ground-excited states. It can be seen that, first, blue Pn is better exposed for interacting with the dopants then black Pn. This is due to the smaller buckling amplitude in the blue Pn and hence better contact interaction (orbital overlaps) between dopant and Pn atoms. The strong donor molecule tetraamino-tetratiofulvalene reduces the energy gap value in both black and blue Pn, thus transforming them into n-semiconductors. The detailed analysis of the electronic band structure [47] shows how the dopants modify the highest occupied and lowest unoccupied energy levels that form the band structure with corresponding changes in observable properties.

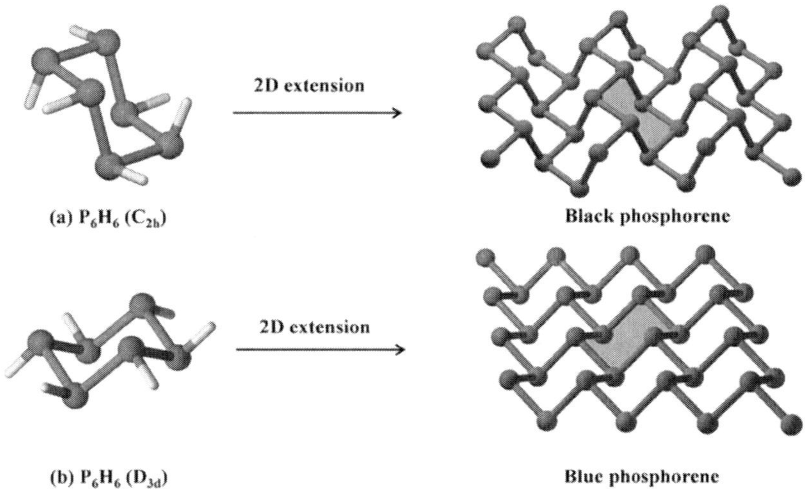

Figure 26. One-to-one mapping between the two PJTE-induced molecular distortions in P_6H_6, e_{1g} and b_{2g}, and the corresponding two buckling forms of phosphorene (Pn), black and blue, respectively: (a) the e_{1g} type distorted configuration of the P_6 ring in P_6H_6 with C_{2h} symmetry is the same as in the local P_6 rings in black Pn; (b) the D_{3d} configuration of the P_6 ring in P_6H_6 due to its PJTE b_{2g} type distortion from the planar regular-hexagonal geometry is the same as the local P_6 ring configuration of blue Pn. The distortion parameters in the molecular and extended systems are very close, in both (a) and (b) cases (the shaded regions in the pictures are shown to ease the visual structural comparisons) (Reprinted with permission from Ref. [47]. Copyright 2016 American Chemical Society).

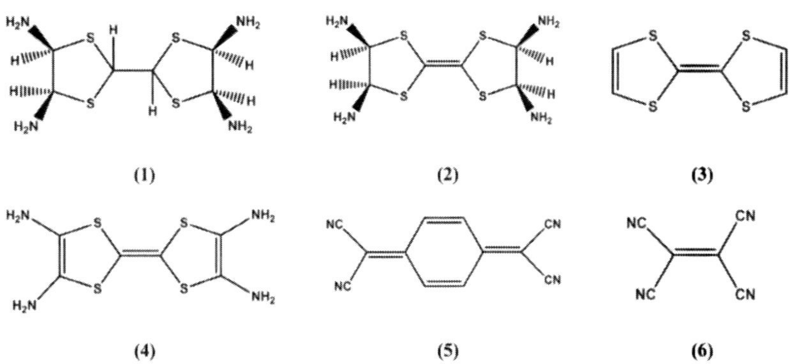

Figure 27. Molecules used as phosphorene dopants: (1) 6H-TTF (hexahydro-tetratiofulvalene), (2) 2H-TTF (dihydro-TTF), (3) TTF, (4) 4NH2-TTF (tetraamino-TTF), (5) TCNQ (tetracyanoquinodimethane), and (6) TCNE (tetracyanoethylene).

4.3.8. 2D Germanene

Similar to silicene, germanene consists of buckled hexa-rings with the buckling amplitude significantly stronger than in silicene; the maximal buckling distance in the $D_{6h} \rightarrow D_{3d}$ distortion of the hexa-ring in Ge_6H_6 is d = 0.61 Å. As in all the other cases of spontaneous distortion of high-symmetry configurations, this buckling is shown to be due to the PJTE induced by the vibronic coupling of the electronic ground state A_{1g} with the low-lying excited state B_{2g} under the nuclear displacements of b_{2g} symmetry [50]. Distinguished from the silicene case the authors suggest another way to influence the germanene properties, namely by hydrogenation to form germanane, in which a large energy gap is opening. This opens a way to control the properties of germanene by further chemical substitutions.

CONCLUSION

The PJTE is shown to be instrumental in revealing the origin of the instability of the planar configuration of heterocyclic systems and their 2D extensions, which leads to puckering or buckling. and, expressed in terms of several parameters of the undistorted system, it allows for manipulation in restoring the planar configuration. The parameters of the PJTE distortions can be extracted from ab initio calculations of the electronic structure of the planar configuration. The most important feature that makes it unstable is the presence of sufficiently low-lying excited states of appropriate symmetry that mix with the ground state under the corresponding lower-symmetry nuclear displacements. In the simplest version, this is described by a relation between three parameters: energy gap Δ, vibronic coupling constant F, and non-vibronic (primary) force constant K_0 (stiffness). The PJTE is the only source of instability of high-symmetry configurations in nondegenerate states (in degenerate states it is additional to the proper Jahn-Teller effect), and as such it includes all particular cases of local interactions that lead to instability (like the often referred to contribution of sp^3 hybridization).

The knowledge of the main PJTE-controlled parameters of instability of planar configurations with respect to distortions opens several opportunities to influence the planarity and related properties of the systems by means of external implications (perturbations), described in this review paper. The qualitative results obtained for the basic structural units of heterocycles remain valid for their extended 2D system, especially with regard to the vulnerable parameters of the PJTE to be targeted in manipulation of the 2D structure. In many cases the local instabilities and distortions remain almost unchanged in the overall 2D system (as, e.g., in black and blue phosphorene, illustrated in Figure 26).

The already realized methods of manipulation of the deviations from the planarity in heterocycles include *targeted* oxidation, reduction, excitation, substitution, and coordination to external atoms or molecular groups. Reviewed examples include metal porphyrins, carbon and silicon cyclic compounds, more complicated tricyclic and polycyclic systems like dithiine containing dithianon, thiophene, thiantrene, anthracene, sumanenes, etc. as well as extended 2D systems of silicene, phosphorene, germanene, and graphitic carbon nitride. In entirety, the PJTE approach to the study of heterocyclic compounds is shown to be a powerful method of exploration of their structure and properties.

References

[1] *2D Materials and Beyond (Synthesis and Doping, Scanned Probe Techniques, 2D Material Devices, Optical Properties of 2D Materials)*, Virtual issue, American Chemical Society Virtual collection, 2017.

[2] 2D Materials: Physics and Device Applications (15 Review Articles), *Chin. Phys. B*, 2017, *26* (3), Special Topic.

[3] Molle, A., Goldberger, J., Houssa, M., Xu, Y., Zhang, S. C. & Akinwande, D. (2017). "Buckled two-dimensional Xene sheets". *Nature Materials*, *16*, 163-169. doi: 10.1038/nmat4802.

[4] Wang, H., Yuan, H., Hong, S., Lib, Y. & Cui, Y. (2015). "Physical and chemical tuning of two-dimensional transition metal dichalcogenids". *Chem. Soc. Rev*, *44*, 2664-2680.

[5] Bersuker, Isaac B. 2006. *The Jahn-Teller Effect*. Cambridge UK: Cambridge University Press;

[6] Bersuker, I. B. & Polinger, V. Z. (1989). *Vibronic Interactions in Molecules and Crystals*. Berlin-Heidelberg-New York: Springer-Verlag; Bersuker, I. B. "Spontaneous Symmetry Breaking in Matter induced by Degeneracy and Pseudodegeneracy". In *Advances in Chemical Physics*, edited by S.A. Rice, A.R. Dinner, 160:159-208. New York: Wiley. 6. Bersuker, I. B. 2013. "Pseudo Jahn-Teller Effect – a two-state paradigm in formation, deformation, and transformation of molecular systems and solids." *Chem. Rev.*, *113*, 1351.

[7] Bersuker, I. B. (2001). "Modern aspects of the Jahn-Teller effect theory and applications to molecular problems". *Chem. Rev.*, *101*, 1067.

[8] Bersuker, Isaac B. (2017). "Manipulation of structure and properties of two-dimensional systems employing the pseudo Jahn-Teller effect." *Flat Chem*, *6*, 11-27.
DOI: 10.1016/j.flatc.2017.10.001.

[9] Gorinchoy, Natalia. (2018). "Pseudo Jahn-Teller effect in puckering and planarization of heterocyclic compounds". *International Journal of Organic Chemistry*, *8*, 142-159. https://doi.org/10.4236/ijoc.2018.81010.

[10] Hermoso, Willian, Ilkhani, Ali R. & Bersuker, Isaac B. (2014). "Pseudo Jahn-Teller origin of instability of planar configurations of hexa-heterocycles $C_4N_2H_4X_2$ (X = H, F, Cl, Br)". *Computational&Theoretical Chemistry*, *1049*, 109-114; see also Ilkhani, Ali R., Hermoso, Willian, & Bersuker, Isaac B. (2015). "Pseudo Jahn-Teller Origin of Instability of Planar Configurations of Hexa-Heterocycles. Application to compounds with C_4X_2 skeletons (X = O, S, Se, Te". *Chemical Physics*, *460*, 75-82.

[11] Liu, Yang, Bersuker, Isaac B., Zou, Wenly, Boggs, James E. (2010). "Pseudo Jahn-Teller versus Renner-Teller Effects in Instability of Linear Molecules". *Chem. Phys.*, *376*, 30-35.

[12] Gorinchoy, Natalia N., Balan, Iolanta I. & Bersuker, Isaac B. (2011). "Jahn-Teller, pseudo Jahn-Teller and Renner-Teller effects in systems with fractional charges". *Computational&.Theoretical Chemistry*, *976* (1-3), 113-119.

[13] Gorinchoy, Natalia N. (2014). "Jahn-Teller and pseudo Jahn-Teller origin of structural distortions of coordinated molecules". *Revue Roumaine de Chimie*, *59*(3-4), 273-280. Gorinchoy, Natalia N. 2016. "Structural distortions of coordinated ketene molecule induced by the pseudo Jahn-Teller effect". *Chem. J. Mold.*, *11*(2), 99-104; http://dx.doi.org/10.19261/cjm.2016.11(2).09.

[14] Gorinchoy N. N. & Bersuker, I. B. (2017). "Pseudo Jahn-Teller effect in control and rationalization of chemical transformations in two-dimensional compounds". *Journal of Physics: Conf. Series*, *833*, 012010. doi :10.1088/1742-6596/833/1/012010.

[15] Schaefer, T., Sebastian, R. & Beaulieu, C. (1991). "Molecular orbital and 1H nuclear magnetic resonance studies of the inversion potentials of thianthrene and thioxanthene". *Canadian Journal of Chemistry*, *69*, 927.

[16] Amato, M. E., Grassi, A., Irgolic, K. J., Pappalardo, G. C. & Radics, L. (1993). "Quantum-mechanical PM3 calculations and high-field proton and carbon-13 NMR studies on chalcanthrenes, $C_{12}H_8XY$ (X, Y = O, S, Se, Te, all combinations)". *Organometallics*, *12*, 775.

[17] Vondele, J. V., Lynden-Bell, R., Meijer, E. J. & Sprik M. (2006). "Density Functional Theory study of tetrathiafulvalene and thianthrene in acetonitrile: structure, dynamics, and redox properties". *J. Phys. Chem. B*, *110*, 3614.

[18] Ito, A., Ino, H., Ichiki, H. & Tanaka, K. (2002). "Tuning Spin-State Preference by Substituents: A Case Study of Thianthrene Dication". *J. Phys. Chem. A*, *106*, 8716.

[19] Lynch, V. M., Simonsen, S.H., Davis, B.E., Martin, G.E., Musmar, M. J., Lam, W. W., Smith, K. (1994). "1,4,6,9-tetraazathianthrene: a

planar thianthrene, $C_8H_4N_4S_2$". *Acta Cryst. C*, *50*, 1470-1472; Mastryukov, V. S., Chen, K. H., Simonsen, S. H., Allinger, N. L. & Boggs, J. E. (1997). "Ab initio and molecular mechanics studies of thianthrene and similar molecules". *J. Mol. Struct.*, *413-414*, 1-12.

[20] Pratik, Saied Md., Chowdhury, C., Bhattacharjee, R., Jahiruddin, Sk. & Datta, A. (2015). "Pseudo Jahn–Teller distortion for a tricyclic carbon sulfide (C_6S_8) and its suppression in S-oxygenated dithiine ($C_4H_4(SO_2)_2$)". *Chemical Physics*, *460*, 101-105.

[21] Chou, Jun-Hong, Rauchfuss Thomas B. & Szczepura, Liza F. (1998). "Studies on the electroactive heterocycles $C_6S_8^{0/2-}$ and $C_6S_6O_2^{0/2-}$ and related metal complexes". *J. Amer. Chem. Soc.*, *120*, 1805-1811.

[22] Benniston, Andrew C., Allen, Ben D., Harriman, Anthony, Llarena, Irantzu, Rostron, James P. & Stewart, Beverly. (2009). "Accessing Molecular Memory via a Disulfide Switch". *New J. Chem.*, *33*, 417-427.

[23] Ilkhani, Ali R., Gorinchoy, Natalia N. & Bersuker, Isaac B. (2015). "Pseudo Jahn-Teller effect in distortion and restoration of planar configurations of tetra-heterocyclic 1,2-diazetes $C_2N_2E_4$, E = H, F, Cl, Br". *Chemical Physics*, *460*, 106-110.

[24] Liu, Y., Wang, Ya & Bersuker, I. B. (2016). "Geometry, electronic structure, and pseudo Jahn-Teller effect in tetrasilacyclobutadiene analogues". *Sci. Rep*, *6*, 23315. doi: 10.1038/srep23315; Liu, Yang, Bersuker, Isaac B., Garcia-Fernandez, Pablo, Boggs, James E. (2012). "Pseudo Jahn-Teller origin of nonplanarity and rectangular ring structure of tetrafluorocyclobutadiene". *J. Phys. Chem. A*, *116*(28), 7564–7570. doi: 10.1021/jp3032836. Liu, Yang, Bersuker, Isaac B. & Boggs, James E. (2013). "Pseudo JahnTeller origin of puckering in $C_4H_4^{2+}$, $Si_4H_4^{2+}$, and $C_4F_4^{2+}$ dications". *Chem. Phys*, *417*, 26-29.

[25] Lin, J. B, Jin, Y., Lopez, S.A., Druckerman, N., Wheeler, S. E. & Houk, K. N. (2017). "Torsional Barriers to Rotation and Planarization in Heterocyclic Oligomers of Value in Organic Electronics". *J. Chem. Theory Comput.*, *13*, 5624-5638.

[26] Gorinchoy, N. N., Arsene, I. & Bersuker, I. B. (2018). "Buckybowl Structure of Sumanenes and Distortions of Thiophenes Induced by the Pseudo Jahn-Teller Effect". *Journal of Physics: Conf. Series, 1148*, 012005. doi :10.1088/1742-6596/1148/1/012005.

[27] Moffat, John B. (1990). *Theoretical Aspects of Heterogeneous Catalysis*. Nelson Canada, Toronto: Van Nostrand Reinhold.

[28] Vessally, E. (2008). "DFT calculations on 1,4-dithiine and S-oxygenated derivatives". *Bull. Chem. Soc. Ethiop.*, *22*(3), 465-468.

[29] Kanagaraj, K., Lin, K., Wu, W., Gao, G., Zhong, Z., Su, D. & Yang, C. (2017). "Chiral Buckybowl Molecules". *Symmetry, 9*, 174; https://doi.org/10.3390/sym9090174.

[30] Priyakumar, U. Deva, & Sastry, G. Narahari. (2001). "Hetero-buckybowls: A Theoretical Study on the Structure, Bowl-to-Bowl Inversion Barrier, Bond Length Alternation, Structure-Inversion Barrier Relationship, Stability, and Synthetic Feasibility". *J. Org. Chem.*, *66*, 6523-30.

[31] Armaković, S., Armaković, S. J., Šetrajčić, J. P., Jaćimovski, S. K. & Holodkov, V. (2014). "Sumanene and its adsorption properties towards CO, CO_2 and NH_3 molecules". *J. Mol. Model.*, *20*, 1-14.

[32] Tahmasebi, E., Biglari, Z. & Shakerzadeh, E. (2017). "Theoretical insight into the impact of sumanene functionalization with BH and NH groups on its ozone addition features". *Vacuum*, 136, 82-90.

[33] Saito, M., Furukawa, S., Kobayashi, J. & Kawashima, T. (2016). "The Chemistry of Heterasumanenes". *Chem. Rec.*, *16*, 64-72.

[34] Pereiray, Z. S. & da Silva, E. Z. (2020). "Spontaneous Symmetry Breaking in Cyclo[18]Carbon". *J. Phys. Chem. A*, DOI: 10.1021/acs.jpca.9b11822.

[35] Wang, Y., Liu, Y. & Zheng, X. (2018). "Pseudo Jahn–Teller origin tracking for symmetry breaking in halogenabenzene: How can a bird fly?" *Int. J. Quantum Chem.*, e25704. https://doi.org/10.1002/qua.25704.

[36] Bersuker, I. B., Stavrov S. S. & Vekhter, B. G. (1979). "Pseudo Jahn-Teller origin of the trigger mechanism of conformational changes in the process of haemoglobin oxygenation". *Biofizica*, *24*, 413.

[37] Bersuker, I. B. & Stavrov, S. S. (1988). "Structure and properties of metalloporphyrins and hemoproteins: the vibronic approach". *Coord. Chem. Revs.*, *88*, 1-68.

[38] Pokhodnya, K., Olson, C., Dai, X., Schulz, D. L., Boudjouk, P., Sergeeva, A. P. & Boldyrev, A. (2011). "Flattening a puckered cyclohexasilane ring by suppression of the pseudo-Jahn–Teller effect". *J. Chem. Phys.*, *134*, 014105. https://doi.org/ 10.1063/ 1.3516179.

[39] Dai, X., Choi, S. B., Braun, C. W., Vaidya, P., Kilina, S., Ugrinov, A., Schulz, D. L. & Boudjouk, P. (2011). "Halide Coordination of Perhalocyclohexasilane Si_6X_{12} (X = Cl or Br)". *Inorg. Chem.*, *50*, 4047-53.

[40] Sergeeva, Alina P. & Boldyrev, Alexander I. (2010). "Flattening a puckered pentasilacyclopentadienide ring by suppression of the pseudo Jahn-Teller effect". *Organometallics*, *29*, 3951-54.

[41] Pokhodnya, K., Anderson, K., Kilina, S. & Boudjouk, P. (2017). "Toward the mechanism of perchlorinated cyclopentasilane (Si_5Cl_{10}) ring flattening in the $[Si_5Cl_{10}{\bullet}2Cl]^{2-}$ dianion". *J. Phys. Chem. A*, *121*, 3494-3500.

[42] Ivanov, Alexander S., Bozhenko, Konstantin V. & Boldyrev, Alexander I. (2012). "On the Suppression Mechanism of the Pseudo-Jahn−Teller Effect in Middle E_6 (E = P, As, Sb) Rings of Triple-Decker Sandwich Complexes". *Inorg. Chem.*, *51*, 8868-72.

[43] Pratik, Saed Md, & Datta, Ayan. (2015). 1,4-Dithiine - Puckered in the Gas-Phase but Planar in Crystals: Role of Cooperativity, *J. Phys. Chem. C*, *119* (27), 15770-76.

[44] Ivanov, A. S., Miller, E., Boldyrev, A. I., Kameoka, Y., Sato, T. & Tanaka, K. (2015). "Pseudo Jahn−Teller Origin of Buckling Distortions in Two-Dimensional Triazine-Based Graphitic Carbon Nitride (g-C_3N_4) Sheets". *Journal of Physical Chemistry C*, *119*, 12008-12015.

[45] Jose, Deepthi, & Datta, Ayan. (2012). "Understanding of the Buckling Distortions in Silicene". *Journal of Physical Chemistry C*, *116*, 24639–48.

[46] Jose, Deepthi, & Datta, Ayan. (2014). "Structures and chemical properties of silicene: unlike grapheme". *Acc. Chem. Res.*, *47*, 593-602. https://doi.org/10.1021/ar400180e.

[47] Chowdhury, Chandra, Jahiruddin, Sheik & Datta, Ayan. (2016). "Pseudo Jahn-Teller Distortion in Two-Dimensional Phosphorus: Origin of Black and Blue Phases of Phosphorene and Band Gap Modulation by Molecular Charge Transfer". *J. Phys. Chem. Lett.*, *7*, 1288-1297.

[48] Qiao, J., Zhou, L. & Ji, W. (2017). "Geometric stability and electronic structure of infinite and finite phosphorus atomic chains". *Chinese Phys. B*, *26*, 036803. DOI:10.1088/1674-1056/26/3/036803.

[49] Likai Li, Jonghwan Kim, Chenhao Jin, Guo Jun Ye, Diana Y. Qiu, Felipe H. Da Jornada, Zhiwen Shi, Long Chen, Zuocheng Zhang, Fangyuan Yang, Kenji Watanab, Takashi Taniguchi, Wencai Ren, Steven G. Louie, XianHui Chen, Yuanbo Zhang, & Feng Wang. (2017). "Direct observation of the layer-dependent electronic structure in phosphorene". *Nat. Nanotechnol.*, *12*, 21-25.

[50] Nijamudheen, A., Bhattacharjee, R., Choudhury, S. & Datta, A. (2015). "Electronic and chemical properties of germanene: the crucial role of buckling". *J. Phys. Chem. C*, *119*, 3802-09. https://doi.org/10.1021/jp511488m.

BIOGRAPHICAL SKETCHES

Isaac B. Bersuker

Affiliation: The University of Texas at Austin, USA

Education: Bachelor in Science from Kishinev University, Moldova, in 1952; PhD in 1957 and Doctor Habilitat in 1964 from Leningrad University (USSR).

Research and Professional Experience: In his Ph. D. dissertation I. B. Bersuker revealed the influence of core polarization on optical transition in atoms (a 1957 paper is still cited in the literature). His main interests afterward were in the Jahn-Teller (JT), pseudo JT (PJT), and Renner-Teller effect theory, and a large variety of their applications. Among his main achievements in this field, he predicted the tunneling splitting in JT systems, proved that the JT and PJT effects are the only source of structural symmetry breaking in polyatomic systems, identified the hidden JT and PJT effects, predicted the coexistence of trigonal and tetragonal JT distortions in T-term JT effect, created the vibronic (PJT) theory of ferroelectricity, discovered the role of spin in spontaneous polarization of crystals and new classes of multiferroics with magnetic-ferroelectric crossover, related vibronic energy level degeneracies to multi-conical intersections and Berry phase, revealed the role of JT and PJT effects in mixed-valence compounds, worked out methods of evaluation JT effect parameters from ultrasonic experiments.

For materials with JT and PJT effects, revealed novel specific properties, including magnetic-dielectric bistability, origin of buckling in 2D systems, giant flexoelectricity, permittivity, electrostriction, and polar nanoregions in paraelectric phases of ferroelectric perovskites.

In a more chemical oriented series of publications, he worked out methods of calculation of PJT induced molecular instabilities, introduced the plasticity effect in stereochemistry, revealed the origin of puckering in molecular systems, showed that bending of linear molecules is induced by the PJT effect (not Renner-Teller effect), worked out a variety of other chemical applications of the JT and PJT effects, including the theory vibronic activation by transition metals in coordination compounds, biological systems, and mixed-valence clusters, explored the role of the PJT effect in hemoglobin oxygenation, revealed the chemical-structural origin of musk fragrance activity, suggested and worked out a method of combined quantum/classical (QM/MM) modeling for large organometallic and metallo-biochemical systems with charge transfer between the QM and MM fragments, and a more "application oriented" Electron-

Conformational method of pharmacophore identification and bioactivity prediction in drug design and toxicology.

Professional Appointments: Head of Laboratory of Quantun Chemistry, Academy of Sciences of Moldova, from 1969 to 1993, Senior researcher and Professor at the Department of Chemistry, the University of Texas at Austin, from 1993.

Honors: Medal of Honor (2004, Moldova), Full Member of the Academy of Science (Moldova, elected in 1988), State Prize Laureate (1980, Moldova), USSR state registered Scientific Discovery N 202 (1979), a series of other medals.

Publications from the Last 3 Years:

1) Bersuker, I. B., The Jahn-Teller and Pseudo Jahn-Teller Effect in Materials Science, *J. Phys. Conf. Series*, 833, 012001 (2017); doi:10.1088/1742-6596/833/1/012001 393.
2) Averkiev, N. S., I. B. Bersuker, V. V.Gudkov, I. V. Zhevstovskikh, M. N.Sarychev, S. Z. herlitsyn, Sh. Yasin, Y. V. Korostelin, V. T. Surikov, Elastic Moduli of Ultrasound Attenuation in Cadmium Selenide doped with Chromium, *J. Appl. Math. Phys.*, 5, 26-30 (2017); http://dx.doi.org/10.4236/jamp.2017.51004.
3) Polinger V., and I. B. Bersuker, Pseudo Jahn-Teller effect in permittivity of ferroelectric perovskites, J. Phys. Conf. Series, 833, 012012 (2017); doi :10.1088/1742-6596/833/1/012012.
4) Averkiev, N. S., I. B. Bersuker, V. V. Gudkov, I. V. Zhevstovskikh, M. N. Sarychev, S. Zherllitsyn, S. Yasin, G. S. Shakurov, V. A. Ulanov and V. T. Surikov, Manifestation of the Jahn-Teller effect in elastic moduli of strontium fluorite crystals doped with chromium ions, *J. Phys. Conf. Series* 833, 012003 (2017); doi :10.1088/1742-6596/833/1/012003.
5) Gorinchoy N. N., and I. B. Bersuker, Pseudo Jahn-Teller effect in control and rationalization of chemical transformations in two-

dimensional compounds, *J. Phys. Conf. Series*, 833, 012010 (2017); doi :10.1088/1742-6596/833/1/012010.

6) Hizhnyakov, V., I. B. Bersuker, and P. Garcia-Fernandez, Eds., Vibronic Coupling and Electron-Phonon Interactions in Molecules and Crystals: XXIII International Symposium on the Jahn-Teller Effect, Foreword, *J. Phys. Conf. Series*, 833, 011001 (2017).
7) Averkiev, N. S., I. B. Bersuker, V. V. Gudkov, I. V. Zhevstovskikh, K. A. Baryshnikov, M. N. Sarychev, S. Zherlitsyn, S. Yasin, and Yu.V. Korostelin, Magnetic field induced tunneling and relaxation between orthogonal configurations in solids and molecular systems, *Phys. Rev. B*, 96, 094431 (2017).
8) Averkiev, Nikita S., Isaac B. Bersuker, Vladimir V. Gudkov,, Irina V. Zhevstovskikh, Maksim N. Sarychev, Sergei Zherlitsyn, Shadi Yasin, Gilman S. Shakurov, Vladimir A. Ulanov, and Vladimir T. Surikov, Acoustic Properties of Crystals with Jahn–Teller Impurities: Elastic Moduli and Relaxation Time. Application to SrF2:Cr2+, *J. Phys. Soc. Jpn*. 86, 114604 (2017); DOI: 10.7566/JPSJ.86.114604.
9) Bersuker, I. B., Manipulation of Structure and Properties of Two-Dimensional System Employing the Pseudo Jahn-Teller effect, *FlatChem*, 6, 11–27 (2017); DOI: 10.1016/j.flatc.2017.10.001.
10) Bersuker, I. B., The Vibronic (Pseudo Jahn-Teller) Theory of Ferroelectricity. Novel Aspects and Applications, *Ferroelectrics*, 536, 1-59 (2018); https://doi.org/10.1080/00150193.2018.1528919.
11) Sarychev, M. N., Irina V. Zhevstovskikh, Nikita S. Averkiev, Isaac B. Bersuker, Vladimir V. Gudkov, V. T. Surikov, Evaluation of the parameters of the Jahn-Teller effect in impurity centers in crystals. Application to the crystal ZnSe:Ni2+, *Russian Solid State Physics*, 61(2), 319-324 (2018).
12) Polinger, V. and I. B. Bersuker, Origin of Polar Nanoregions and Relaxor properties of ferroelectrics, *Phys. Rev. B*, 98, 214102 (2018). DOI:10.1103/PhysRevB.98.214102.
13) Averkiev, N. S., I. B. Bersuker, V. V. Gudkov, I. V. Zhevstovskikh, M. N. Sarychev, S. Zherlitsyn, S. Yasin, Yu. V. Korostelin and V. T. Surikov, Jahn-Teller effect problems via ultrasonic experiments.

Application to the impurity crystal, CdSe:Cr, *J. Phys. Conf. Series*, 1148 (2018) 012008; doi :10.1088/Journal of 1742-6596/ 1148/1/012008.

14) Gorinchoy, N. N., I. Arsene, and I. B. Bersuker, Buckybowl Structure of Sumanenes and Distortions of Thiophenes Induced by the Pseudo Jahn-Teller Effect, *J. Phys. Conf. Series*, 1148 (2018) 012005; doi:10.1088/1742-6596/1148/1/012005.

15) Сарычев, М.Н., И.В. Жевстовских, Н.С. Аверкиев, И. Б. Берсукер, В.В. Гудков, В.Т. Суриков, Определение параметров эффекта Яна-Теллера в примесных центрах из ультразвуковых экспериментов: приложение к кристаллу ZnSe:Ni^{2+}, *Физика Твёрдого Тела*, ФТТ, 61(2), 319-324 (2019); doi:10.21883/ FTT.2019.02.47132.243

16) Baryshnikov, Kirill, Nikita Averkiev, Isaac Bersuker, Vladimir Gudkov, Irina Zhevstovskikh, Maksim Sarychev, Sergei Zherlitsyn, Shadi Yasin, and Yuriy Korostelin, Interplay between relaxation and resonance in ultrasound attenuation by the cubic crystal ZnSe:Cr, *Phys. Status Solidi* B 2019,1800635; doi: 10.1002/pssb.201800635

17) Wang, Ya, Yang Liu, and Isaac B. Bersuker, Sudden polarization and zwitterions formation as a pseudo Jahn-Teller effect: A new insight in photochemistry of alkenes, *Phys. Chem. Chem. Phys.*, 21, 10677 – 10692 (2019), doi: 10.1039/C9CP01023H.

18) Аверкиев, Н.С., I.B. Bersuker, В.В. Гудков, И.В. Жевстовских, М.Н. Сарычев, S. Zherlitsyn, S. Yasin, Ю.В. Коростелин, В.Т. *Суриков, Определение параметров эффекта Яна-Теллера в примесных кристаллах с помощью ультразвуковых исследований*, 2019, ЖЭТФ, 2019, том 156, вып.1(7), стр. 87–97.

19) Averkiev, N. S., I. B. Bersuker, V. V. Gudkov,, I. V. Zhevstovskikh, M. N. Sarychev, S. Zherlitsyn, S. Yasin, Yu. V. Korostelin, and V. T. Surikov, Ultrasonic Determination of the Jahn–Teller Effect Parameters in Impurity-Containing Crystals, *J. Exp. Teor. Phys*. 129, 72-80 (2019).

20) Bersuker, I. B., The Jahn-Teller and Pseudo Jahn-Teller Effect in Materials Science, in *Recent Studies in Materials Science*, Ed. Patrick R. Lind, Nova Science Publishers, 2019, pp. 1-95.
21) Averkiev, Nikita S., Isaac B. Bersuker, Vladimir V. Gudkov, Irina V. Zhevstovskikh, Maksim N. Sarychev, Sergei Zherlitsyn, Shadi Yasin, Gilman S. Shakurov, Vladimir A. Ulanov, and Vladimir T. Surikov, Manifestation of the Jahn-Teller Effect in Elasic Moduli of Cubic Crystals. Application to Strotium Fluorite Doped with Chromium Ions, in *Fluorite: Structure, Chemistry, and Applications,* Ed. Moritz van Asten, Nova Science Publishers, 2019, pp 110-159.

Natalia N. Gorinchoy

Affiliation: Institute of Chemistry, Academy of Sciences of Moldova, Republic of Moldova.

Education: Bachelor in Science from Irkutsk State University (USSR), in 1973; PhD in 1978 from Leningrad State University (USSR).

Research and Professional Experience: In her Ph. D. dissertation N. N. Gorinchoy carried out ab initio calculations of potential energy surfaces for a number of chemical reactions, and worked out a method and software for revealing localized orbitals in the MC SCF theory. Her main interests afterward were in the Jahn-Teller (JT) and pseudo JT (PJT) effect theory, and a large variety of their applications. Among her main achievements in this field, she worked out a method and software for numerical calculations of orbital vibronic coupling constants, revealed the origin of distortions of high-symmetry nuclear configurations of a number of molecular systems, including systems with fractional charges, and methods of enhancement and suppression of the PJTE in manipulation molecular structures.

Professional Appointments: Researcher, Senior researcher and Leading researcher at the Laboratory of Quantum Chemistry, Institute of Chemistry, Academy of Sciences of Moldova, from 1980 to 2010; Head of Laboratory of Quantum Chemistry, Institute of Chemistry, Academy of Sciences of Moldova, from 2010.

Publications from the Last 3 Years:

1) Gorinchoy, N. N. Pseudo Jahn-Teller effect in puckering and planarization of heterocyclic compounds, *International Journal of Organic Chemistry*, 8, 142-159 (2018); doi:10.4236/ ijoc.2018.81010.
2) Gorinchoy N. N., and I. B. Bersuker, Pseudo Jahn-Teller effect in control and rationalization of chemical transformations in two-dimensional compounds, *J. Phys. Conf. Series,* 833, 012010 (2017); doi :10.1088/1742-6596/833/1/012010.
3) Gorinchoy, N. N., I. Arsene, and I. B. Bersuker, Buckybowl Structure of Sumanenes and Distortions of Thiophenes Induced by the Pseudo Jahn-Teller Effect, *J. Phys. Conf. Series,* 1148, 012005 (2018); doi:10.1088/1742-6596/1148/1/012005.
4) 4) Osipov, I., Gorbachev, M. Yu., and Gorinchoy, N. N., Proton-induced accelerated decay of the fungicide, vinclozolin, on TiO2 surface under solar irradiation: Experimental and DFT study, *Journal of environmental science and health, Part B*, 53 (4), 222–228, (2018); doi: 10.1080/03601234.2017.1410406.
5) Gorbachev, M. Yu., Gorinchoy, N. N., and Arsene, I., Surface tension of near-boiling liquids with intramolecular $\pi\sigma^*$-hyperconjugation: DFT calculations, *Journal of Molecular Liquids*, 237, 108–112, (2017); doi: 10.1016/j.molliq.2017.04.069.
6) Kanarovskii, E. Yu., Yaltychenko, O. V., and Gorinchoy, N. N., Kinetics of Antioxidant Activity of α-Tocopherol and Some of its Homologues: Part 1. Review: Theoretical Model, *Surface Engineering and Applied Electrochemistry,* 54(5), 481–497, (2018);. ISSN 1068-3755, *Allerton Press, Inc.*, 2018.

7) Arsene, I. and Gorinchioy, N., DFT study of the entire reaction cycle of H_2O_2 decomposition and O_2 generation catalyzed by Fenton reagent, *Chemistry Journal of Moldova*, 14(1), 88-97 (2019); doi: 10.19261/cjm.2018.543.

In: Heterocycles
Editor: Jonas M. Danielsen

ISBN: 978-1-53617-804-3

Chapter 4

ORGANIC AZIDES: PRECURSORS OF TRIAZENES

Enrique Barragan, PhD and Alejandro Bugarin*, PhD
Department of Chemistry and Physics,
Florida Gulf Coast University, Fort Meyers, FL, US

ABSTRACT

Organic azides are among the most versatile functional groups in organic chemistry. In addition to the widely known traditional reactivity pathways of the azido group itself, the additional capability of azides to form key reactive intermediates in a straightforward manner greatly expands their versatility and synthetical utility. One of the most salient reactive intermediates that can be readily accessed from organic azides is the triazene group. This functional group consists of three contiguous nitrogen atoms with a double bond between the first and second nitrogen. These adducts were recently divided, by Bugarin, in three categories: linear, cyclic, and π-conjugated triazenes. Once these adducts/ intermediates have been formed, they can be immediately transformed into a variety of functional groups. In the present chapter, we'll discuss the reaction of azides with Grignard or organolithium reagents either in

* Corresponding Author's E-mail: abugarin@fgcu.edu.

an inter- or intramolecular fashion to form linear or cyclic triazene intermediates, respectively. Through these reactions, azides can be converted into important functionalities such as; amines, amides, and nitrogen heterocycles, which in turn can function as synthetic building blocks for more complex chemical structures. Additionally, we will present known applications that make use of the power of azides' reactivity for the synthesis of prominent chemotherapeutic drugs bearing a triazene core structure.

Last but not least, throughout the latter part of the chapter, we will explore recently reported azide reactivity with *N*-heterocyclic carbenes under basic conditions, which allows their conversion into aldehydes, ketones, ethers, thioethers, (hetero)biaryls, and aryl amides. This novel reactivity permitted the conversion of organic azides into useful functional groups under mild, transition-metal-free conditions, for example; (hetero)biaryls and aryl amides formation under a choice of photochemical or sunlight irradiation.

Introduction

There is no question that the azide functional group is among the most versatile moieties in organic chemistry. Throughout the years, these compounds have found numerous applications such as; dipolar cycloadditions, aniline synthesis, heterocycles' precursors, the core structure of pharmaceuticals, and many more (Bräse et al. 2005, 5189). In this Chapter, however we will focus on an often-overlooked type of azide reactivity, their conversion into a closely-related triazene group and reactions that are possible to achieve by transformation of these triazene intermediates.

The triazene functional group consists of three contiguous nitrogen atoms, with a double bond between the first and second nitrogen and connected to carbon atoms on both ends (Patil and Bugarin 2016, 860). As shown in Figure 1, there are three main types of triazenes: linear, cyclic, and π-conjugated triazenes. Both linear 1 and cyclic triazenes 2 are prepared from azides following a similar synthetic approach and possess a similar reactivity (Haley and Kimball 2002, 3340-49); hence, the first part of this chapter will be dedicated to discuss their preparation, reactivity, and importance in chemotherapeutic drug preparation. On the other hand, π-conjugated triazenes 3 are a relatively unexplored family of molecules for which new reactions have been emerging in very recent years (Patil and Bugarin 2016, 861-869). These fascinating compounds, their synthesis from azides as well as their reactivity, will be covered in their own section on the latter part of this chapter.

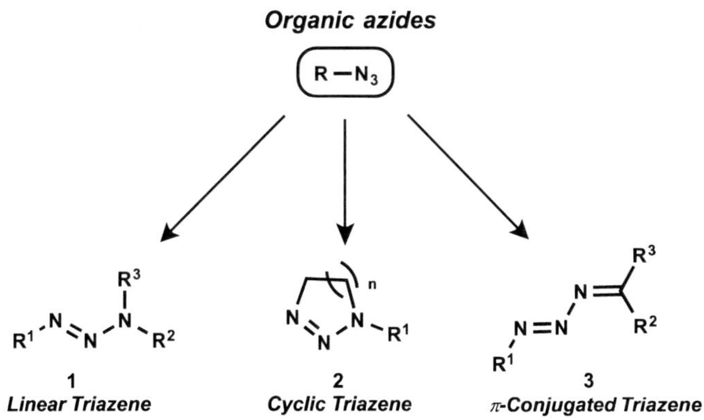

Figure 1. Classes of triazenes accessible from organic azides.

CONVERSION OF AZIDES TO TRIAZENES

Transformation of azides to linear triazenes is a fairly straightforward process. As illustrated in Figure 2, the organic azide 4 reacts with an alkyl carbanion, the source of which can be either a Grignard or an organolithium reagent, to form intermediate 5, which in turns furnishes the desired triazene 6 after an acidic work-up. These transformations are very efficient and deliver their adduct in good to excellent yields (Kimball and Haley 2002, 3339).

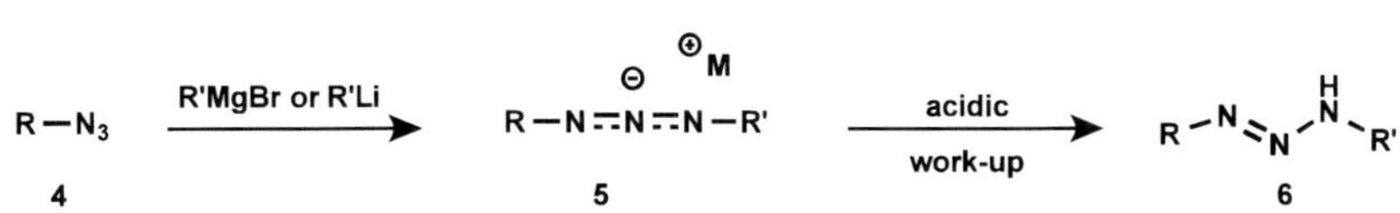

Figure 2. Synthesis of linear triazenes from azides.

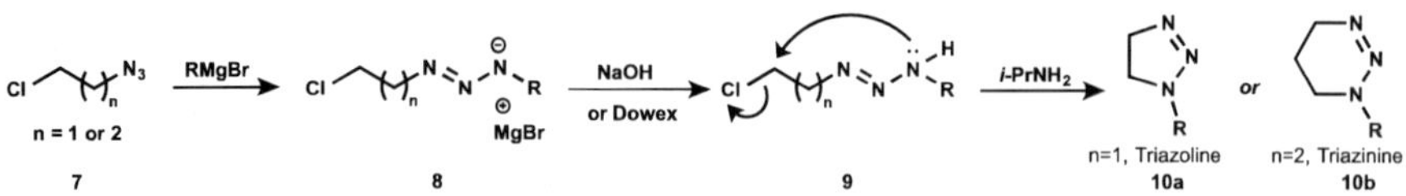

Figure 3. Synthesis of triazolines and triazinines from azides.

Regarding the use of organic azides to generate cyclic triazenes, although less abundant in literature than their linear counterparts, some reports about their synthesis and reactions have appeared throughout the years. As is to be expected, these cyclic species can also be formed from an azide precursor in an analogous manner to the linear type, with the particularity that their formation typically involves an intramolecular step (Patil and Bugarin 2016, 861).

One such approach that utilizes azides for the synthesis of five and six-membered cyclic triazenes, also named triazolines and triazinines, respectively, was reported by Michejda and coworkers (Smith et al. 1992, 6448). In this method, represented in Figure 3, a derivative of either chloroethyl or chloropropyl azide 7 was reacted with a Grignard reagent to form the ionic triazene magnesium salt 8, after treatment of salt 8 with sodium hydroxide or Dowex resin, neutral triazene 9 was released. This lineal triazene intermediate spontaneously underwent intramolecular cyclization under the reaction conditions, displacing the chloride ion. Finally, treatment with a mild base such as isopropyl amine neutralizes the resulting protonated triazene and furnishes the resulting neutral triazoline 10a (Smith et al. 1992, 6453) or triazinine 10b (Schmidt et al. 1997, 8663), respectively.

As noticed from the presented reactions, organic azides can be easily converted into both linear and cyclic triazenes by reaction with organometallic reagents. However, conversion of azides into π-conjugated triazenes can be classified in their own category, therefore, the transformations of azides into π-conjugated triazenes will be discussed in a further section within this Chapter. In the following sections, representative reactions of azides that involve linear and cyclic triazene intermediates will be presented, as well as highlighting the biological activity and importance of these molecules in the preparation of chemotherapeutic agents.

REACTIONS OF AZIDES *VIA* LINEAR AND CYCLIC TRIAZENE INTERMEDIATES

Azides in conjunction with organometallics have been used to access amines, common synthetic intermediates that can also serve to access nitrogen-containing heterocycles, through the intermediacy of linear triazene species (Trost and Pearson 1981, 2483-4). The top of Figure 4 illustrates this approach, in which azidomethyl phenyl sulfide is used in combination with either a Grignard reagent directly or a combination of organolithium compound 11 and magnesium bromide, to produce a linear triazene intermediate 12 that, without the need for purification, is then directly hydrolyzed using aqueous potassium hydroxide to the desired aryl amine 13.

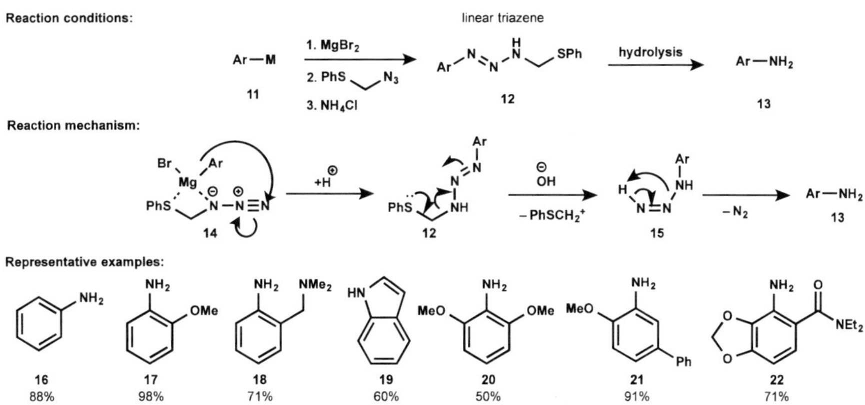

Figure 4. Synthesis of amines from azides through linear triazenes.

The sulfur present in azido methyl phenyl sulfide seems to facilitate the attack from the organometallic reagent to the azide *via* coordination to magnesium as shown in structure 14 in the reaction mechanism shown in the middle of Figure 4, thus providing a pathway for base-catalyzed decomposition to the amine 13 (Trost and Pearson 1981, 2484).

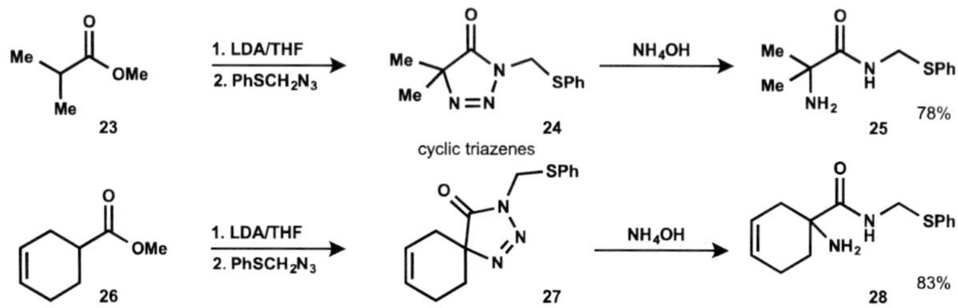

Figure 5. Synthesis of amines from azides through cyclic triazenes.

Some advantages of this method include its applicability to sterically congested substrates, as evidenced by the products 18 and 22, in which the amide group was successfully introduced in sterically hindered positions. The ortho regioselectivity of this approach which can be observed in products 17, 18, and 21 offers a complementary route to a standard electrophilic substitution reaction. Formation of heterocycles is also realized as evidenced by the formation of indole 19, Additionally, formation of the anthranilic acid derivative 22 showcases the potential utility to be used in the generation of substituted arynes derived from anthranilic acids (Trost and Pearson 1981, 2484).

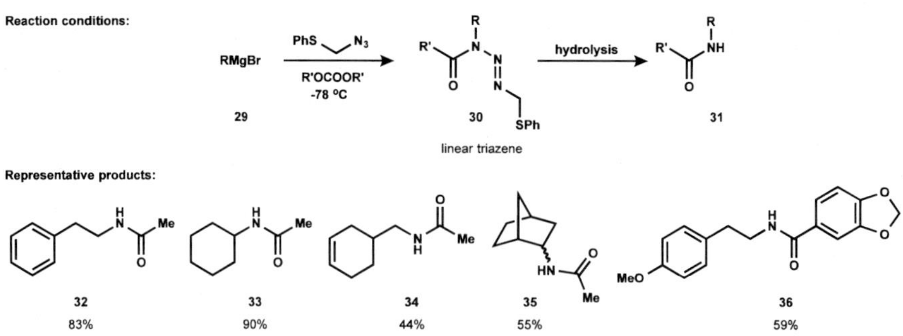

Figure 6. Synthesis of amides (protected amines) from azides.

Moreover, this approach has also been found to work with enolates derived from α,α-disubstituted esters. In these cases, the reaction proceeds through a cyclic triazene intermediate 24 that readily hydrolyzes in an analogous manner to the linear triazene approach, furnishing α-aminoamides 25 in a straightforward fashion, as shown in Figure 5 (Trost and Pearson 1981, 2485).

A variant of this method was also reported by Trost in which azido methyl phenyl azide is reacted with aliphatic organometallic species 29 and acid anhydrides at low temperature, followed by hydrolysis of triazene intermediate 30, generating *N*-substituted alkyl amides 31, as shown in Figure 6 (Trost and Pearson 1983, 1055). From the generated amides, the corresponding alkyl amines can then be produced through amide cleavage or amide reductions (not shown) (Trost and Pearson 1983, 1055). Products 32 to 36 demonstrate the structural diversity that is available from this approach. Furthermore, this report highlights the versatility of the azide moiety by showing how it allows access to different functional groups while tolerating a variety of substituents, *via* formation of triazene intermediates, through simple adjustments to the reaction conditions.

Applications of Organic Azides in the Synthesis of Chemotherapeutic Agents

One of the most important applications of linear and cyclic triazenes that have emerged in recent years is the development of new chemotherapeutic agents to treat a variety of tumors. Two of the most salient chemotherapy drugs are shown in Figure 7. Dacarbazine 37 is a highly active linear triazene that is extensively used for the treatment of Hodgkin's lymphoma and malignant carcinoma (Vaughan 1991, 159). On the other hand, temozolomide 38 is a cyclic triazene that is commonly prescribed for the treatment of glioblastoma, a particularly aggressive type of brain tumor (Hegi et al. 2005, 997).

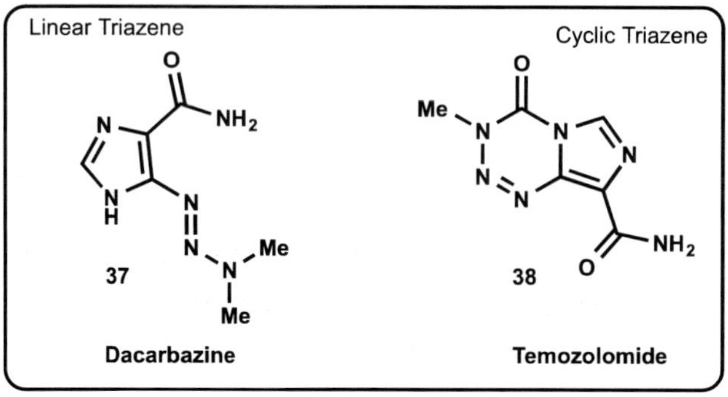

Figure 7. Prominent chemotherapeutic triazene drugs.

Aside from these prominent examples, research on these molecules continues to find more triazenes with antitumor activity. The key for the effectiveness of these compounds lies on their metabolic activity. Under physiological conditions, linear and cyclic triazenes decompose generating highly reactive diazonium species that can alkylate DNA (Kleihues et al. 1976, 2192). Although this reactivity may make some of these compounds mutagenic or carcinogenic (Krüger et al. 1971, 529), there are cases like those of the drugs previously mentioned in Figure 7 that possess low toxicity compared to their potency against cancer cells (Gilbert et al. 2002, 261).

Some types of triazenes that have attracted particular interest for their potential antitumor applications are 1,3-dialkyltriazenes 40, 1,3,3-trialkyltriazenes 43, and derivatives of 1,3-dialkyl-3-acyltriazenes, represented in Figure 8 as structures 41 and 44 to 46 (Vaughan 1991, 179). Of particular interest for the purposes of this chapter is the fact that all of these triazene subclasses are accessible from azide precursors (Vaughan 1991, 180).

As discussed previously and represented in Figure 8, in order to synthesize 1,3-dialkyltriazenes, an alkyl azide 39 must be reacted with an organometallic reagent. Once obtained, the dialkyltriazene 40 can be further alkylated by treating it with a strong base, generally potassium *tert*-butoxide or potassium hydride in the presence of 18-crown-6, followed by addition of methyl iodide, affording the desired trialkyltriazene 43 (Vaughan 1991, 180).

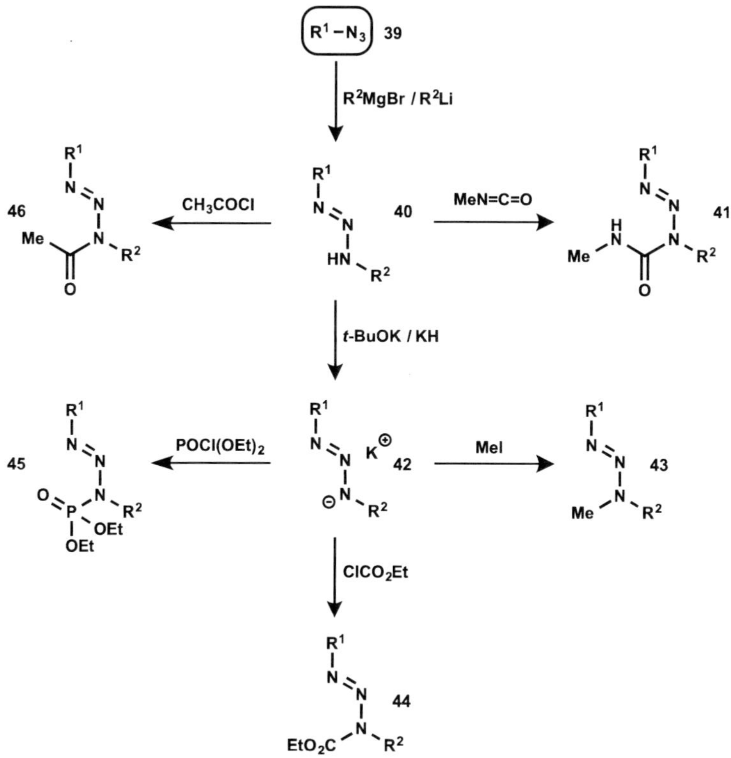

Figure 8. Conversion of azides to triazenes with chemotherapeutic potential.

Figure 8 also shows routes for the obtention of other triazene derivatives of interest. If the dialkyltriazene 40 is reacted with acetyl chloride, selective acetylation occurs at N-3 to furnish 3-acetyl-1,3-dialkyltriazenes 46. Following the same principle, acylating the triazene anion with ethyl chloroformate or diethyl chlorophosphate produces the ester 44 and diethyl phosphonate 45 derivatives, respectively. Finally, reaction of the dialkyltriazene 40 with methyl isocyanate results in the direct synthesis of the carbamoyltriazene 41 (Vaughan 1991, 180).

As illustrated in Figure 8, a wide variety of triazene subclasses are accessible from the same azide precursor in a few straightforward steps. This remarkable versatility highlights the potential that the azide functional group has and most certainly additional applications will be elucidated, in the near future, during the search for new anticancer treatments.

Reactivity of Azides with *N*-Heterocyclic Carbenes

The final type of triazene intermediate that can be accessed through organic azides is the π-conjugated triazene. This particular kind of triazene contains a double bond between the third nitrogen of the triazene group and the carbon atom attached to it (Patil and Bugarin 2016, 861). Since this configuration possesses conjugated double bonds over four atoms, a situation analogous to the structure of 1,3-butadiene, this particular family of compounds is also sometimes referred also as; 1,2,3-triazabutadienes (Kimani and Jewett 2015, 4051).

Although these compounds were first synthesized over fifty years ago (Winberg and Coffman 1965, 2776), it wasn't until recent years that a straightforward and convenient method for their preparation utilizing azides was developed (Patil, White, and Bugarin 2014, 4826). In 2005, Bielawski employed organic azides 47 to react with readily available large imidazolium salts 48. Sodium hydride or potassium *tert*-butoxide was used to deprotonate those imidazolium salts to *in situ* generate reactive *N*-heterocyclic carbenes (NHCs). This clever combination furnished their respective π-conjugated triazenes 49 in a single step with good to excellent yields (Khramov and Bielawski 2005, 4958), as illustrated in Figure 9.

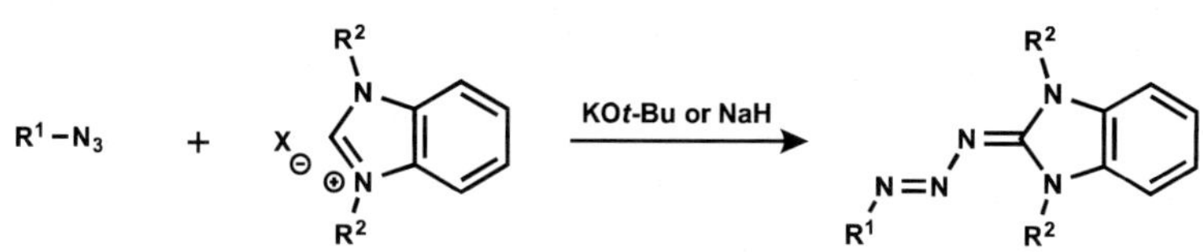

Figure 9. Reaction of azides with NHCs to form π-conjugated triazenes.

This approach has been applied to react azides with a variety of NHC precursors to obtain π-conjugated triazenes with diverse structural and electronic properties. For instance, as shown in Figure 10, by reacting organic azides 47 with benzimidazolium salts 48 under basic conditions, bulky triazenes 49 were obtained, which were shown to exhibit bond alteration patterns characteristic of donor-acceptor molecules when analyzed by X-ray crystallography (Patil and Bugarin 2014, 221). It was also determined that the electronics of the azide play a fundamental role in determining the stability of the resulting triazene product, since triazenes derived from electron-deficient aromatic azides have higher thermal stability compared to triazenes derived from more electron-rich azides (Khramov and Bielawski 2007, 9407-14).

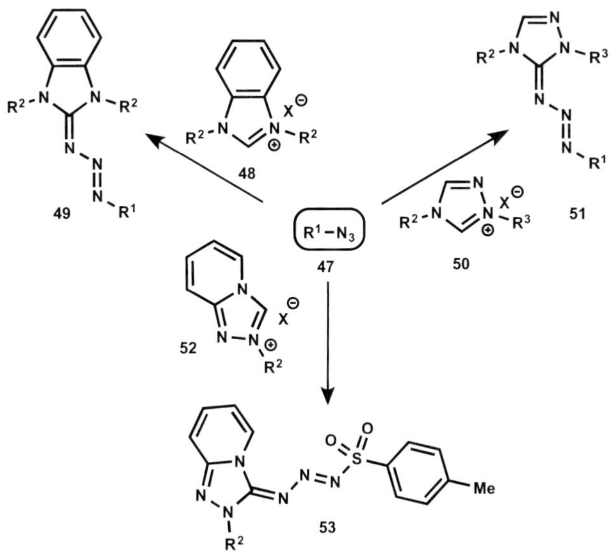

Figure 10. Conversion of organic azides into diverse π-conjugated triazenes.

In 2013, a similar method was used by Katritzky and coworkers to react azides with both symmetrically and asymmetrically-substituted triazolium salts 50 to generate the corresponding triazenes 51, as illustrated in Figure 10. Moreover, azides were also employed in tandem with anionic triazolium-derived NHC's to generate novel triazene salts, *an advance in the area since they are water soluble*. Most of the triazenes produced exhibited exceptional thermal stability, with a few exceptions whose rapid decomposition at high temperatures could make them potential blowing agents for energetic formulations. X-ray analysis of these compounds also evidenced donor-acceptor properties like their benzimidazole-derived counterparts. This was further confirmed by reacting these triazenes with nucleophiles like MeI (iodomethane), with the results showing extremely high selectivity towards the terminal nitrogen, suggesting an increase of electron density on that nitrogen due to the push-pull nature of these molecules (Jishkariani et al. 2013, 3350-51).

More recently, Wei and coworkers reported a simple method to prepare triazenes derived from sulfonyl azides, as shown in the bottom reaction of Figure 10. In their approach, they reacted sulfonyl azides with diverse substituents in the aromatic ring with benzotriazolium salts 52 and potassium *tert*-butoxide in acetone at 20 °C for 2 hours, under an argon atmosphere, yielding the corresponding π-conjugated triazenes 53 in good to excellent yields. Interestingly, one of the triazenes obtained showed a linear and selective fluorescence response in the presence of Fe^{3+} ion over a concentration range relevant to drinking water maximum permissible limits. This observation suggests the potential to use this or very similar triazenes as sensors to selectively detect Fe^{3+} ions (Yi et al. 2019, 1059-61).

All previous examples highlight straightforward transformations of azides into different kinds of π-conjugated triazenes, compounds with emerging useful applications in material and biological sciences. In addition, in the following sections of this chapter, we will explore how the conversion of azides into triazenes can expand traditional azide reactivity by way of these molecules functioning as valuable reaction intermediates that can allow access to several key functional groups, later utilized in synthetic organic and material chemistry.

Substitution and Oxidation Reactions of Azides

The first reaction we will discuss is the oxidation of organic azides to aldehydes or ketones via π-conjugated triazene intermediates. Our research group reported this unprecedented reaction in 2017, it was first discovered while investigating the thermal stability of the triazene derived from benzyl azide 54 and dimethylimidazolium 55 in deuterated DMSO. The main product from this transformation was guanidine 57, derived from nitrogen gas extrusion. Benzaldehyde 58 was also identified as one of the products after triazene consumption at 120 °C, for 2 h (Barragan and Bugarin 2017, 1499), as represented in Figure 11.

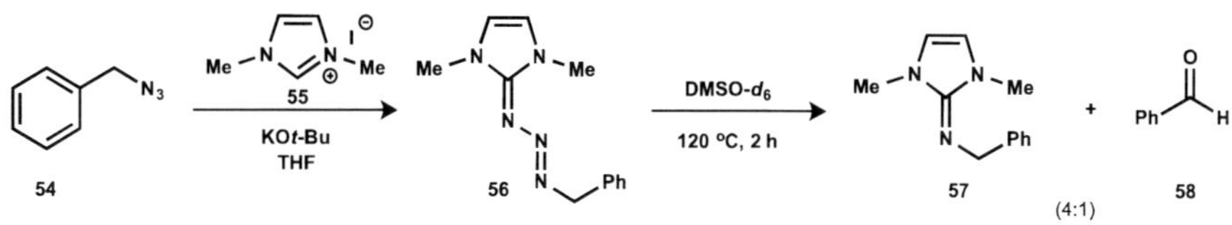

Figure 11. Initial observation of azide oxidation reaction.

The observed oxidation reaction was further optimized by incorporating *p*-toluenesulfonic acid (TsOH) as a Brønsted acid catalyst, which activates intermediates to favor the oxidation reaction to occur at room temperature. With optimized conditions in hand, it becomes possible to carry out the oxidation reaction directly from the azide in a single pot without the need to isolate triazene intermediates (Barragan and Bugarin 2017, 1500). Figure 12 illustrates some examples of the results obtained while exploring the scope of this novel transformation by using different benzylic azides as substrates.

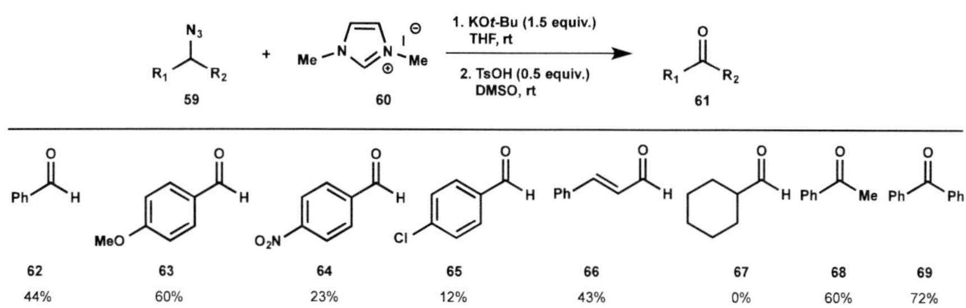

Figure 12. Azide oxidation to carbonyl compounds.

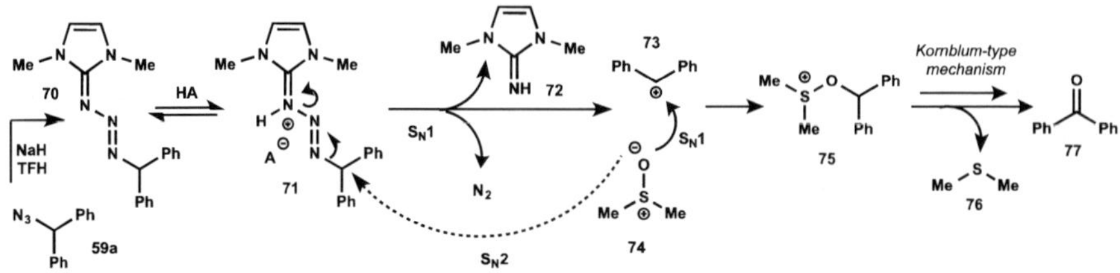

Figure 13. Proposed mechanism for the azide oxidation reaction.

The data obtained revealed the following trend: A decrease in reaction yields in accordance with decreasing stability of a carbocation intermediate, with no product 67 obtained for the primary substrate and the best yield obtained for compound 69 resulting from the secondary, dibenzylic substrate, see Figure 12. This observation suggests that the reaction proceeds through an S_N1 reaction mechanism with a positively-charged intermediate involved. Figure 13 shows the proposed mechanism for the title transformation, which begins with protonation of triazene intermediate 70, followed by fragmentation to a carbocation 73, guanidine byproduct 72, and loss of molecular nitrogen. Finally, the carbocation 73 is captured by DMSO 74 and the resulting alkoxysulfonium intermediate 75 is converted to the corresponding carbonyl product 77 and the observed dimethyl sulfide 76 (Barragan and Bugarin 2017, 1501).

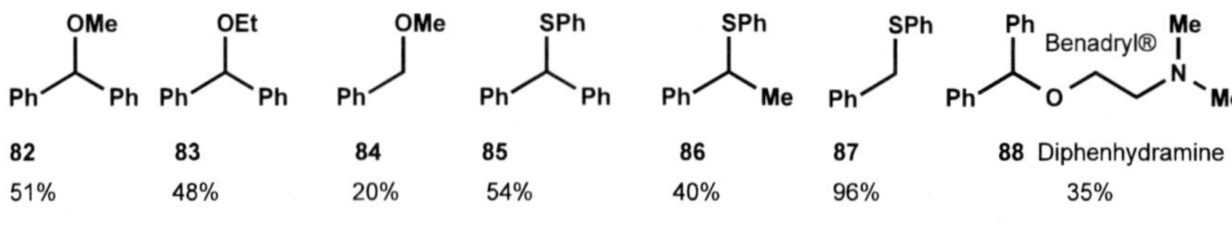

Figure 14. Substitution reaction of azides via triazene intermediates.

Since the proposed mechanism involves a nucleophilic substitution step, different nucleophiles can be used in place of DMSO to produce substitution products. This was confirmed in our study by utilizing alcohols and thiols to generate the corresponding ethers 82-84 and sulfides 85-87 from azides, as shown in Figure 14. Interestingly, strongly acidic thiols such as thiophenol derivatives are able to self-catalyze the reaction, furnishing sulfide products 85-87 without the need for external Brønsted acid. This approach allows the use of benzhydryl azide as a precursor to the drug diphenhydramine (Benadryl®) 88, a widely used antihistaminic agent (Barragan and Bugarin 2017, 1501). These examples highlight and expand on the known versatility of organic azides by showing how, with slight variations in reaction conditions and via the intermediacy of a π-conjugated triazene species, they can be transformed into an array of different significant and useful functional groups. Those freshly prepared carbonyl compound, ethers and sulfides can then serve as building blocks for the design and construction of more complex organic frameworks.

Photochemical Conversion of Azides into (Hetero)biaryls

Cross-coupling reactions are among the most prominent in organic chemistry since they allow access to biaryl systems which are present in numerous biologically-active compounds (Hassan et al. 2002, 1359). Traditionally, these transformations are carried out using Pd catalysts. Relevant examples of such transformations are: Suzuki-Miyaura (Miyaura 2002, 12), Negishi (Negishi et al. 1977, 1821-22), Stille (Stille 1986, 508) and Hiyama (Hatanaka and Hiyama 1988, 918) reactions. Although highly efficient, these traditional methods often involve stoichiometric amounts of toxic organometallic species such as organotin and organozinc compounds, which require specialized procedures for their synthesis, purification, recovery and disposal (Friedman et al. 2012, 13-21). Additionally, the need to use high temperatures limits functional group tolerance as well as long reaction times also increases operation cost (Dach et al. 2012, 1703). In response to these shortcomings, alternative methods that reduce the use of toxic metals have been developed in the past few years. These complementary methods typically involve the use of highly reactive species such as aryliodonium salts (Storr and Greaney 2013, 1410-13), aryl-diazotates (Hofmann et al. 2017, 9647-54) and aryldiazonium salts (Ghosh et al. 2016, 1566-75), in combination with a catalytic amount of a transition metal catalyst. A summary of those methods is depicted in Figure 15.

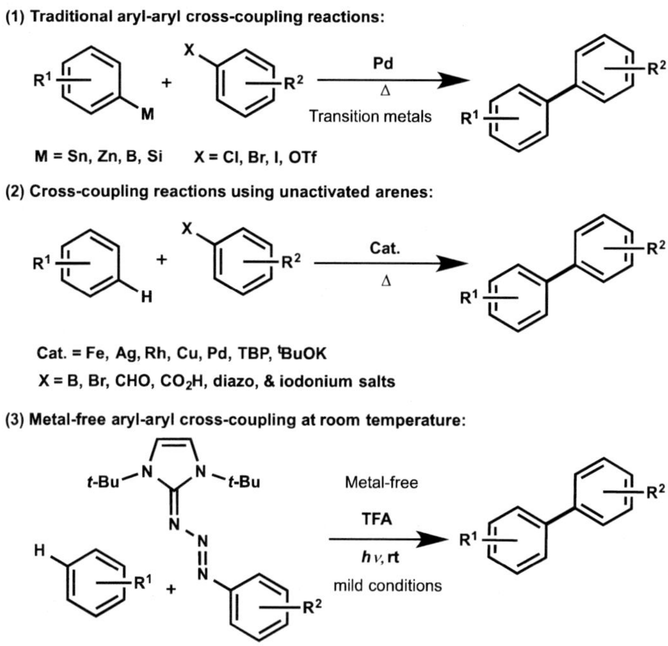

Figure 15. Aryl-aryl cross-coupling reactions.

Out of these alternative methods for cross-coupling reactions, the use of aryldiazonium salts is one of particular interest since it generates molecular nitrogen as a clean byproduct as well as for the capability of diazonium ions to be converted to either aryl cations or aryl radicals by tuning the reaction conditions (Pazo-Llorente et al. 2006, 2201). Nevertheless, the approach is limited by the risk incurred in handling diazonium salts in larger scales due to their known instability and propensity to cause explosions (Agrawal and Hodgson 2007, 148), resulting in a need to find milder alternatives.

Our group recently reported one such approach that utilizes aromatic azides as substrates to produce benzamides *via* the formation of a π-conjugated triazene intermediate (Barragan et al. 2019, 152). This approach is represented in Figure 16, where it can be seen that the aryl azide 89 is first converted to a π-conjugated triazene intermediate 91 by reaction with an imidazolium salt 90 under basic conditions. Followed by treatment of this intermediate 91 with trifluoroacetic acid in an arene solvent and UV irradiation for 2 h at room temperature to produce the cross-coupling biaryl products 92. Representative examples 93-101 of the substrate scope of this transformation are also depicted in Figure 16.

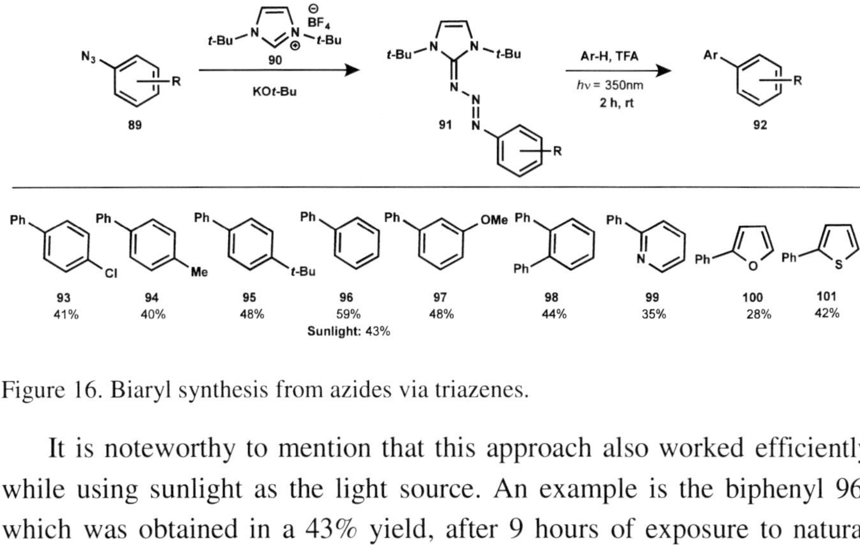

Figure 16. Biaryl synthesis from azides via triazenes.

It is noteworthy to mention that this approach also worked efficiently while using sunlight as the light source. An example is the biphenyl 96, which was obtained in a 43% yield, after 9 hours of exposure to natural sunlight in a quartz tube. This observation increases the potential applicability of the method from a sustainability perspective. It is also important to note that this approach also allows the synthesis of various heterobiaryls, as shown in Figure 16, which often can constitute a challenge (Hofmann et al. 2017, 9647).

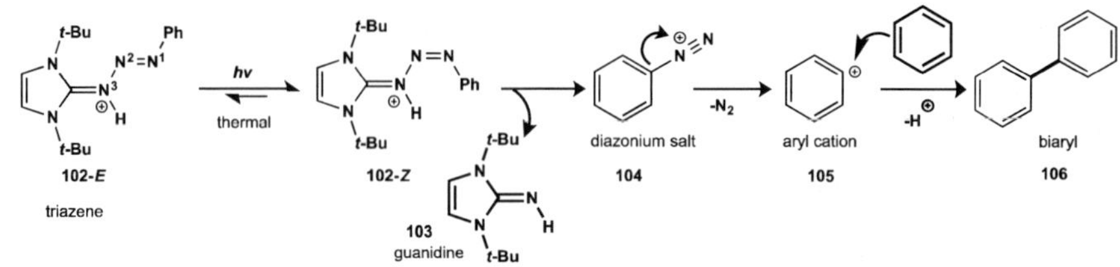

Figure 17. Proposed mechanism for the cross-coupling reaction.

By using reactivity data and DFT calculations, we have elucidated a plausible mechanism through which this reaction might occur. As illustrated in Figure 17, we believe that the reaction mechanism begins when the aromatic azide-derived triazene is protonated in nitrogen 3 and is then photo-isomerized from the more stable *E*-isomer 102-*E* to the more reactive *Z*-form 102-*Z*. Our calculations suggest that protonation at nitrogen 3 is the only pathway conducive to reaction, with protonation at nitrogen 1 resulting in the equilibrium favoring the starting material (Barragan et al. 2019, 156-7). Once the excited state of the *Z*-isomer of the triazene intermediate relaxes back to the ground state, fragmentation occurs to generate a guanidine byproduct 103 and a diazonium salt 104 *in situ,* which quickly loses molecular nitrogen to furnish the corresponding aryl cation 105. Finally, the aryl cation is captured by the arene and the biaryl product 106 is formed after rearomatization (Barragan et al. 2019, 158-9).

The synthetic utility of our approach was demonstrated by the three-step synthesis of biologically-active compound 111 in Figure 18, which possesses immunosuppressant properties (Storr and Greaney 2013, 1413).

In summary, this method is yet another example of how the versatility of azides can be greatly expanded from the traditional reactions by their conversion to π-conjugated triazenes which, acting as intermediates, allow for the straightforward transformation of the azide moiety into functional groups as diverse as aldehydes, ketones, ethers, sulfides and biaryls by tuning the reaction conditions and substrate structure.

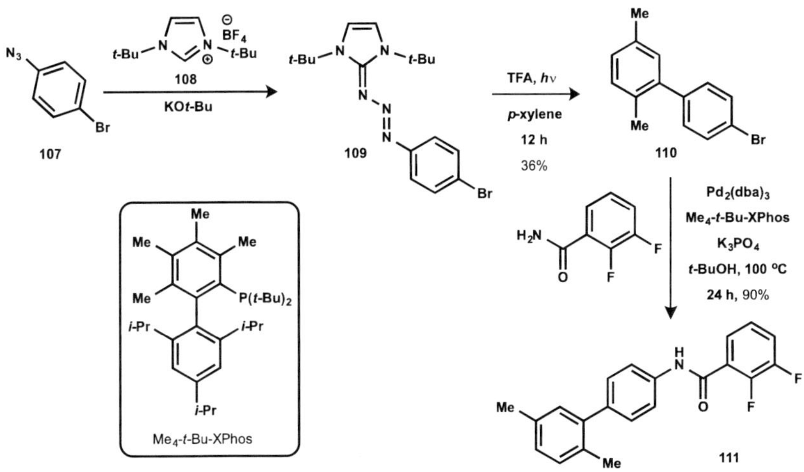

Figure 18. Synthesis of immunosuppressant.

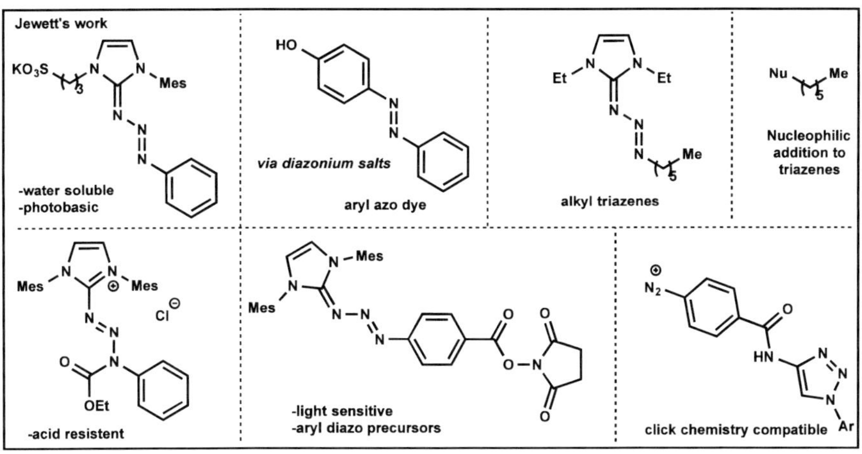

Figure 19. Jewett's current work.

Furthermore, and before closing this chapter, we would be remiss if we failed to mention a potentially exciting new azide reaction through π-conjugate triazenes. While studying the previously discussed aryl-aryl cross-coupling reaction, we have observed the formation of arylamides under similar conditions by varying the solvent used. Preliminary studies suggest that by a slight variation in reaction conditions, arylamides can be obtained in good to excellent yields in a metal-free mild fashion (to be published soon).

Last but not least, a major player in the chemistry of π-conjugated triazenes (1,2,3-triazabutadiene) is Professor Jewett (University of Arizona). This far, he has reported several triazabutadienes prepared from organic azides, see Figure 19. For instance, Jewett has described a water-soluble triazabutadienes (He, Kimani, and Jewett 2015, 9764) with tunable pH that, under acidic conditions, formed a diazonium species that subsequently, when trapped with phenol, furnished the known aryl azo dye (Kimani and Jewett 2015, 4051). Jewett also reported uses of diazonium ions from triazabutadienes (He, Kimani, and Jewett 2017, 1767), triazenes as alkylating agents (Knyazeva et al. 2017, 2700), light activated triazabutadienes (Jensen et al. 2016, 2216), acid resistant triazabutadienes (Guzman et al. 2016, 2220), coumarin triazabutadienes for protein labeling (Shadmehr et al. 2018, 2550),and triazabutadienes that are compatible with click chemistry, dipolar cycloadditions, conditions (Cornali et al. 2016, 4948), as illustrated on Figure 19. Notably, all these approaches used relative bulky carbenes, to introduce stability. Nonetheless, all reports are very clever and simple strategies to introduce functionality in these useful and important molecules, while increasing the number of applications of such molecules.

OUTLOOK AND FUTURE DIRECTION

In summary, as we have discussed throughout this chapter, the azide functional group has an outstanding potential for biology, synthetic and material chemistry due to its remarkable versatility. In this chapter, we have presented how the known azide versatility can be expanded further by producing the triazene functional group as means to access many different applications.

Reactions involving the triazene group, although not as thoroughly explored as many other reaction pathways for the azide moiety have, nonetheless, garnered increasing interest in recent years due to their simplicity and synthetic potential. We have analyzed how by reacting alkyl azides with organometallic compounds, linear and cyclic triazenes can be readily obtained. These compounds can be easily hydrolyzed to yield useful synthetic building blocks, including amines and amides. Furthermore, dialkyltriazenes can be easily alkylated or acylated further to obtain trialkyl and dialkylaryltriazenes, compounds known for their cytotoxic action through DNA alkylation, which makes them a highly studied family of compounds for the development of new chemotherapeutic drugs.

Not limited to the previous examples, azides can also be readily converted to π-conjugated triazenes, a kind of molecules that has not been properly explored until very recent years. Through these fascinating intermediates, benzylic azides can undergo acid-catalyzed oxidation reactions to carbonyl compounds or be subject to nucleophilic substitution to furnish products such as ethers and sulfides under mild conditions. Additionally, through these π-conjugated triazene intermediates, aryl azides can undergo aryl-aryl cross coupling reactions to yield biaryls or heterobiaryls under UV irradiation or natural sunlight in a transition metal-free route. Moreover, as current studies indicate, it might also be possible to obtain aryl amides in a similar transition-metal-free, mild and straightforward fashion by a slight procedure variation.

In conclusion, azide reactions involving triazenes are a practical and versatile alternative to traditional methods for the obtention of diverse functional groups and, due to the momentum of recent developments, we are certain that it is a field that will continue to grow in the future. We are looking forward and excited to see forthcoming reports from Professor Jewett and others. Regarding to our future plans, we will continue investigating this kind of versatile molecules and hope to release more exciting transformations and applications in the near future.

REFERENCES

Agrawal, J. P. and Hodgson, R. D. 2007. *Organic Chemistry of Explosives*. United States: John Wiley & Sons, Ltd.

Barragan, E. and Bugarin, A. 2017. "π-Conjugated Triazenes: Intermediates That Undergo Oxidation and Substitution Reactions." *The Journal of Organic Chemistry* 82. 1499-1506.

Barragan, E., Noonikara Poyil, A., Yang, C. H., Wang, H. and Bugarin, A. 2019. "Metal-free cross-coupling of π-conjugated triazenes with unactivated arenes *via* photoactivation." *Organic Chemistry Frontiers* 6. 152-161.

Bräse, S., Gil, C., Knepper, K. and Zimmermann, V. "Organic Azides: An Exploding Diversity of a Unique Class of Compounds." *Angewandte Chemie International Edition* 44. 5188-5240.

Cornali, Brandon M., Flora W. Kimani, and John C. Jewett. 2016. "Cu-Click Compatible Triazabutadienes to Expand the Scope of Aryl Diazonium Ion Chemistry." *Organic Letters*. 18. 4948-4950.

Dach, R., Song, J. J., Roschangar, F., Samstag, W., Senanayake, C. H. 2012. "The Eight Criteria Defining a Good Chemical Manufacturing Process." *Organic Process Research & Development* 16. 1697-1706.

Friedman, D., Masciangioli, T. and Olson, S. 2012. *The Role of the Chemical Sciences in Finding Alternatives to Critical Resources: A Workshop Summary*. Wahington, DC: The National Academies Press.

Ghosh I., Marzo, L., Das, A., Shaikh, R. and König, B. 2016. "Visible Light Mediated Photoredox Catalytic Arylation Reactions." *Accounts of Chemical Research* 49. 1566-1577.

Gilbert, M. R., Friedman, H. S., Kuttesch, J. F., Prados, M. D., Olson, J. J., Reaman, G. H. and Zaknoen, S. L. "A phase II study of temozolomide in patients with newly diagnosed supratentorial malignant glioma before radiation therapy." 2002. *Neuro-Oncology* 4. 261-267.

Guzman, Lindsay E., Flora W. Kimani, and John C. Jewett. 2016. "Protecting Triazabutadienes to Afford Acid Resistance." *Chem Bio Chem* 17. 2220-2222.

Hassan, J., Sévignon, M., Gozzi, C., Schulz, E. and Lemaire, M. 2002. "Aryl-Aryl Bond Formation One Century after the Discovery of the Ullmann Reaction." *Chemical Reviews* 102. 1359-1469.

Hatanaka Y. and Hiyama T. 1988. "Cross-Coupling of Organosilanes with Organic Halides Mediated by Palladium Catalyst and Tris(diethylamino)sulfonium Difluorotrimethylsilicate." *The Journal of Organic Chemistry* 53. 918-920.

Hegi, M. E., Diserens, A. C., Gorlia, T., Harmou, M. F., Tribolet, N., Weller, M., Kros, J. M., Hainfellner, J. A., Mason, W., Mariani, L., Bromberg, J. E. C., Hau, P., Mirimanoff, R. O., Cairncross, J. G., Janzer, R. C. and Stupp, R. 2005. "*MGMT* Gene Silencing and Benefit from Temozolomide in Glioblastoma." *The New England Journal of Medicine* 352. 997-1003.

He, J., Kimani, F. W. and Jewett, J. C. 2015. "A Photobasic Functional Group." *Journal of the American Chemical Society* 137. 9764-9767.

He, J., Kimani, F. W. and Jewett, J. C. 2017. Rapid in Situ Generation of Benzene Diazonium Ions under Basic Aqueous Conditions from Bench-Stable Triazabutadienes. *Synlett* 28. 1767-1770.

Hofmann, J., Gans, E., Clark, T. and Heinrich M. R. 2017. "Radical Arylation of Anilines and Pyrroles via Aryldiazotates." *Chemistry A European Journal* 23. 9647-9656.

Jensen, Stephanie M., Flora W. Kimani, and John C. Jewett. 2016. "Light-Activated Triazabutadienes for the Modification of a Viral Surface." *ChemBioChem* 17. 2216-2219.

Jishkariani, D., Hall, C. D., Demircan, A., Tomlin, B. J., Steel, P. J. and Katritzky, A. R. 2013. "Push-Pull Triazenes Derived from 1-(Benzylideneamino)- and 1-(Sulfonimido)-azolylidenes." *The Journal of Organic Chemistry* 78. 3349-3354.

Khramov, D. M. and Bielawski, C. W. 2005. "Triazene formation *via* reaction of imidazole-2-ylidenes with azides" *Chemical Communications*. 4958-4960.

Khramov, D. M. and Bielawski, C. W. 2007. "Donor-Acceptor Triazenes: Synthesis, Characterization, and Study of Their Electronic and

Thermal Properties." *The Journal of Organic Chemistry* 72. 9407-9417.

Kimani, F. W. and Jewett, J. C. 2015. "Water-Soluble Triazabutadienes that Release Diazonium Species upon Protonation under Physiologically Relevant Conditions." *Angewandte Chemie International Edition* 54. 4051-4054.

Kimball, D. B. and Haley, M. M. 2002. "Triazenes: A Versatile Tool in Organic Synthesis." *Angewandte Chemie International Edition* 41. 3338-3351.

Kleihues, P., Kolar, G. F. and Margison, G. P. 1976. "Interaction of the Carcinogen 3,3-Dimethyl-1-phenyltriazene with Nucleic Acids of Various Rat Tissues and the Effect of a Protein-free Diet." *Cancer Research* 36. 2189-2193.

Knyazeva, D. C., Kimani, F. W., Blanche, J. L. and Jewett, J. C. 2017. "Hexyl triazabutadiene as a potent alkylating agent," *Tetrahedron Letters* 58. 2700-2702.

Krüger, F. W., Preussmann, R. and Niepelt, N. 1971. Mechanism of carcinogenesis with 1-aryl-3,3-dialkyl-triazenes-III: *In vivo* methylation of RNA and DNA with 1-phenyl-3,3-(^{14}c)-dimethyltriazene." *Biochemical Pharmacology* 20. 529-533.

Miyaura, N. 2002. "Organoboron Compounds" In *Cross-Coupling Reactions: A Practical Guide*, edited by Norio Miyaura, 11-59. Springer-Verlag Berlin Heidelberg.

Negishi, E., King, A. O., Okukado, N. 1977. "Selective Carbon-Carbon Bond Formation via Transition Metal Catalysis. A Highly Selective Synthesis of Unsymmetrical Biaryls and Diarylmethanes by the Nickel- or Palladium- Catalyzed Reaction of Aryl- and Benzylzinc Derivatives with Aryl Halides." *The Journal of Organic Chemistry* 42. 1821-1823.

Patil, S. and Bugarin, A. 2016. "Fifty Years of π-Conjugated Triazenes." *European Journal of Organic Chemistry*. 860-870.

Patil, S. and Bugarin, A. 2014. "Crystal Structures of (E)-1,3-dimethyl-2-(3(-nitrophenyl)triaz-2-enylidene)-2,3-dihydro-1H-imidazole." *Acta Crystallographica Section E.*, 70. 221-227.

Patil, S., White, K. and Bugarin, A. 2014. "Novel Triazene Dyes from N-Heterocyclic Carbenes and Azides: Synthesis, Stability, and Spectroscopic Properties" *Tetrahedron Lett.* 55. 4826-4829.

Pazo-Llorente, R., Maskill, H., Bravo-Diaz, C. and Gonzalez-Romero E. 2006. "Dediazonation of 4-Nitrobenzenediazonium Ions in Acidic MeOH/H_2O Mixtures: Role of Acidity and MeOH Concentration on the Formation of Transient Diazo Ethers that Initiate Homolytic Dediazoniation." *European Journal of Organic Chemistry*. 2201-2209.

Schmidt, B. F., Snyder, E. J., Carroll, R. M., Farnsworth, D. W., Michejda, C. J. and Smith, R. H. 1997. "Triazinines: Synthesis and Proteolytic Decomposition of a New Class of Cyclic Triazenes." *The Journal of Organic Chemistry* 62. 8660-8665.

Smith, R. H., Wladkowski, B. D., Herling, J. A., Pfaltzgraff, T. D., Taylor, J. E., Thompson, E. J., Pruski, B., Klose, J. R. and Michejda, C. J. 1992. "Novel Triazolines from the Base-Catalyzed Hydrolysis of 1,3-Dialkyl-3-acyltriazenes." *The Journal of Organic Chemistry* 57. 6448-6454.

Shadmehr, M., Davis, Garrett J., Mehari, B. T., Jensen, S. M. and Jewett, J. C. 2018. "Coumarin triazabutadienes for fluorescent labeling of proteins." *ChemBioChem* 19. 2550-2552.

Stille, J. K. 1986. "The Palladium-Catalyzed Cross-Coupling Reactions of Organotin Reagents with Organic Electrophiles." *Angewandte Chemie International Edition* 25. 508-524.

Storr, T. E. and Greaney, M. F. 2013. "Palladium-Catalyzed Arylation of Simple Arenes with Iodonium Salts." *Organic Letters* 15. 1410-1413.

Trost, B. M. and Pearson, W. H. 1981. "Azidomethyl Phenyl Sulfide. A Synthon for NH_2^+." *Journal of the American Chemical Society* 103. 2483-2485.

Trost, B. M. and Pearson, W. H. 1983. "Sulfur Activation of Azides towards addition of Organometallics. Amination of Aliphatic Carbanions." *Journal of the American Chemical Society* 105. 1054-1056.

Vaughan, K. 1990. "Triazenes" In *The Chemistry of Antitumour Agents*, edited by Derry E. V. Wilman, 159-186. Springer, Dordrecht.

Winberg, H. E. and Coffman, D. D. 1965. "Chemistry of Peraminoethylenes." *Journal of the American Chemical Society* 87. (12), 2776-2777.

Yi, D., Wu, Y., Chen, C., Wang, L., Wang, Q., Pu, L. and Wei, S. 2019. "Preparation of pyrido(1,2-*c*)(1,2,4)triazole-based π-conjugated triazene as a Fe^{3+} ion fluorescent sensor." *Chinese Chemical Letters* 30. 1059-1062.

INDEX

B

C

D

E

M

N

O

P

T

U

V

W

X

Π